International Journal of Agricultu‌ ty

Urban agriculture: diverse activities and benefits for city society

Routledge
Taylor & Francis Group

LONDON AND NEW YORK

VOLUME 8 ISSUE 1&2 2010

International Journal of Agricultural Sustainability

Urban agriculture: diverse activities and benefits for city society

First published 2010
by Earthscan

Published 2014 by Routledge
2 Park Square, Milton Park, Abingdon, Oxon OX14 4RN

Simultaneously published in the USA and Canada
by Routledge
711 Third Avenue, New York, NY, 10017, USA

Routledge is an imprint of the Taylor & Francis Group,
an informa business

First issued in paperback 2016

The International Journal of Agricultural Sustainability is
indexed in Thomson ISI Current Contents®/Agriculture,
Biology & Environmental Sciences and Science Citation
Index Expanded™

Abstracting Services which cover this title include
Elsevier Scopus, CSA Sustainability Science Abstracts
(Cambridge Scientific Abstracts) and CAB Abstracts.

ISBN13: 978-1-84971-124-1 (hbk)
ISBN13: 978-1-138-98657-2 (pbk)

The *International Journal of Agricultural Sustainability* (IJAS) is a cross-disciplinary, peer-reviewed journal dedicated to advancing the understanding of sustainability in agricultural and food systems. IJAS publishes both theoretical developments and critical appraisals of new evidence on what is not sustainable about current or past agricultural and food systems, as well as on transitions towards agricultural and rural sustainability at farm, community, regional, national and international levels, and through food supply chains. It is committed to clear and consistent use of language and logic, and the use of appropriate evidence to substantiate empirical statements. IJAS increases knowledge on what technologies and processes are contributing to agricultural sustainability, what policies, institutions and economic structures are preventing or promoting sustainability, and what relevant lessons should be learned.

Topics covered include: agriculture–environment interactions, agriculture and rural economy interactions, institutional and policy issues, technology development, and food and consumer issues.

Notes for Contributors

1. SUBMISSION PROCEDURE

Submissions for IJAS should be sent as Word files to ijas@essex.ac.uk. All articles will be peer-reviewed before acceptance and authors should include full contact details for three potential reviewers. Submissions may also be sent to:

The IJAS journal, c/o Prof Jules Pretty
Department of Biological Sciences
University of Essex
Colchester
CO4 3SQ
UK.

Authors should keep a copy of their articles and illustrations. While the Editors and Publisher will take all possible care of material submitted to the Journal, they cannot be held responsible for the loss of or damage to any material in their possession. The final decision on acceptance will be made by the Editor.

2. TYPES OF SUBMISSION

IJAS accepts: Research articles (maximum 7000 words), Review articles (maximum 7000 words), Commentaries (maximum 1000 words), Book Reviews and Letters to the Editor.

3. LANGUAGE AND STYLE

Articles should be in English and should be written and arranged in a style that is succinct and easy for readers to understand. Authors who are unable to submit their articles in English should contact the Editors so that any alternatives may be considered. Illustrations should be used to aid the clarity of the article; do not include several versions of similar illustrations, or closely-related diagrams, unless each is making a distinct point.

4. MANUSCRIPT PREPARATION AND LAYOUT

The first page of the manuscript should contain the full title of the article, the author(s) names without qualifications or titles, and the affiliations and full address of each author. The precise postal address, telephone and fax numbers and email address of the author to whom correspondence should be addressed should also be included. The second page should contain an abstract of the article (c. 200 words) and a key word list (up to 6 words). The abstract should précis the article, giving a clear indication of its conclusions. Within the body of the article, headings and sub-headings should be used so that the article is easy to follow.

TABLES AND SCHEMA

Authors should aim to present table data as succinctly as possible and tables should not duplicate data that are available elsewhere in the article. All tables must have a caption.

SYMBOLS, ABBREVIATIONS AND CONVENTIONS

Please use SI (Systeme Internationale) units. Whenever an acronym or abbreviation is used, ensure that it is spelled out in full the first time it appears. If unusual symbols (e.g. Greek letters) are used in the paper, authors are asked to send a hard copy of the article, with such symbols highlighted.

REFERENCES

References should be presented in 'author/date' style in the text and collected in alphabetical order at the end of the article. All references in the reference list should appear in the text. Each reference must include full details of the work referred to, including paper or chapter titles and opening and closing page numbers

Journals:
Lieblein, G., Ostergaard, E. and Francis, C. (2004) Becoming an agroecologist through action education, *International Journal of Agricultural Sustainability* 4, 134–162.

Books:
Jones, J.H. (2003) Grazing impacts on the nutrient cycle. In: Grant, L. (ed.) *Grazing Systems*. London: Earthscan.
Grant, L. (2003) *Grazing Systems*. London: Earthscan.

Proceedings:
Wentworth, S. and Hu, F. (2004) Agricultural policy development in the mid-west. In: T.G. Smith (ed.), *Proceedings of the 8th Annual Meeting of the Association for Agricultural Development*, London: James & James.

NOTES

Notes should be kept to a minimum and will appear as endnotes. Indicate endnotes with a superscript number in the text, and include the text at the end of the article. Do not use the footnote/endnote commands in word processing software for either references or notes.

ILLUSTRATIONS

Illustrations must relate clearly to the section in which they appear and should be referred to in the text as Figure 1, Figure 2 etc. Each illustration requires a caption. They should be submitted in a form ready for reproduction – no redrawing or re-lettering will be carried out by the Publisher. Illustrations should be supplied as tiffs, jpegs or editable eps files; the filename must include the corresponding authors surname and figure number. Note that figures and graphs must be comprehensible in black-and-white – use patterns, not colours, to differentiate sections.

5. PROOFS AND OFFPRINTS

The corresponding author will receive proofs for correction; these should be returned to Earthscan within 72 hours of receipt. The corresponding author will receive a PDF file of the published article. Offprints are available for purchase and must be ordered prior to publication; an order form will be provided for this purpose.

6. COPYRIGHT

Submission of an article to the journal is taken to imply that it represents original work, not under consideration for publication elsewhere. Authors will be asked to transfer the copyright of their articles to the Publisher. Copyright covers the distribution of the material in all forms including but not limited to reprints, photographic reproductions and microfilm. It is the responsibility of the author(s) of each article to collect any permissions and acknowledgements necessary for the article to be published prior to submission to the Journal.

GUEST EDITORIAL: Challenging, multidimensional agriculture in cities

Craig J. Pearson*

Melbourne Sustainable Society Institute, The University of Melbourne, VIC 3010, Australia

This special volume showcases urban agriculture – its potential to provide multiple benefits and its challenges, generically grouped as social, economic and environmental goods and services (see Pearson et al., pp. 7–19; Leeuwen et al., pp. 20–25). The majority of the world's population now lives in cities, and this proportion, and the size of the cities themselves, will increase dramatically over the next 20 years. It is known that city greenspace and local food production bring multiple benefits, but these are mostly not recorded by researchers and ignored by public policy makers. Consequently, the retention of greenspace and the encouragement of food production are usually based on ad hoc decisions by elected local councils without the benefit of access to independently researched information.

'Urban agriculture' is not a single entity. It encompasses residual, often peri-urban broadacre farmland, small 'community gardens', personally managed allotments, home gardens, portions of parks that were previously planted entirely with amenity species, fruit trees along roadside reserves, greenhouses, green roofs and green walls. These diverse activities, scales and locales share many opportunities and problems. These are distinctively different from the conditions encountered by rural agriculture (Eriksen-Hamel and Danso, pp. 86–93).

This volume focuses on the social and environmental impacts of urban agriculture. Socially, it may play a central role in community development (Seymoar et al., pp. 26–39; Karanja et al., pp. 40–53) and provide unique benefits (Leeuwen et al.; Sumner et al., pp. 54–61). Environmentally, in addition to providing food, urban agriculture has impacts on landscape aesthetics, air and water (e.g. Merson et al., pp. 72–85;

Eriksen-Hamel and Danso). However, one volume cannot be comprehensive: we do not include greenhouse food production because there is a robust scientific literature on greenhouse food, and greenhouses are not uniquely urban although some of their side effects are (e.g. light pollution).

The authors bring their distinct perspectives as agriculturists, environmental scientists and social scientists: planners, landscape architects and community development specialists dealing with issues such as resource use and social cohesion. If this volume seems 'heavier' in social sciences, that may reflect the view that the core of sustainability is social, while it must also address economic and biophysical concerns. Though comprehensive economic analyses are important, none were submitted to this special issue. The future challenges will be huge, especially in the face of the possibility of the failure of the food supply systems to megacities. As with the negative impacts of climate change or diminishing resources (e.g. oil), the impact of diminishing greenspace or urban food system failure will hurt society's poor first and hardest.

What about the future? It seems to me that the paramount issue regarding the sustainability of urban agriculture, underlying many papers and central in Condon et al. (pp. 104–115), is the social one of how to protect urban land for greenspace and food production. This challenge involves governance, legal issues and the question of how to create sustainable market economies that, frankly, have rarely been achieved other than within underemployed, near-subsistence communities or with niche-branded value-added exceptions where the food, wine and recreation are bundled. More broadly, while the papers in this issue cover a broad spectrum from

*Email: c.pearson@unimelb.edu.au

INTERNATIONAL JOURNAL OF AGRICULTURAL SUSTAINABILITY 8 (1&2) 2010

PAGES 3–4, doi:10.3763/ijas.2009.c5008 © 2010 Earthscan. ISSN: 1473-5903 (print), 1747-762X (online). www.earthscan.co.uk/journals/ijas

high-income cities to the poorest city precincts in developing countries, they could all support the proposal for four things to achieve a sustainable future for urban agriculture (Redwood, pp. 5–6). These needs are first, to protect open space within cities; second, to accept that agriculture is a legitimate land use and to legalize it and if necessary take action, e.g. public liability insurance to protect it; third, to promote and maintain food markets; and fourth, to establish food policy councils to coordinate municipal responses to urban food security.

Again, these are social and governance issues. Several papers identify researchable issues such as documenting the quantity and nutritional value of fresh local food, impacts on health, and mixed-use design such as green walls, and the desirability of co-researching with property developers to overcome social, economic and biophysical hurdles to add both monetary value and social capital to communities. One contribution from urban planners is more radical: while stressing the need for greater attention to issues such as social equity and diet, Knight and Riggs (pp. 116–126) call for a new paradigm of 'nourishing urbanism'. Clearly, urban agriculture is, if you will pardon the pun, a fertile area for research that is arguably more challenging and multidimensional, and likely to have greater community impact, than research in rural agriculture.

COMMENTARY: Food price volatility and the urban poor

Mark Redwood*

Urban Poverty and Environment Program, International Development Research Centre (IDRC), 150 Kent St., P. O. Box 8500, Ottawa, Ontario, Canada

In 2006, the United Nations marked the first year in which more than 50 per cent of humanity is living in cities and towns. Given the short time scale in which urbanization has occurred, it is not an exaggeration to say that this is arguably the most significant human migration in history. More so, it is a migration with profound consequences for both society and economies worldwide.

Urbanization has two faces – on one side, the promise and potential of cities as engines of growth that have driven the creation of wealth and pulled many out of poverty (China and India are prime examples of this). On the other hand, there is a more desperate scenario of cities whose growth has outpaced the ability to govern them, leading to failing infrastructure, a lack of basic services for the poor, dense slum areas and highly polluted landscapes.

One underexamined aspect of cities is where farming – generally considered a rural activity – intersects with the city. 'Urban' and 'Agriculture' have only been recently paired, with most planners and agriculturalists preferring the convenience of their separation. In reality, agriculture is a fundamental part of the landscape of most cities. Thus, given the current volatility in commodity markets, urban farming merits renewed attention.

The causes of the recent food price spike were a perfect storm of high oil prices, constrained supply, the legacy of certain trade policies, biofuel demand, changing diets in rapidly growing economies and, above all, fear and speculation – two profoundly powerful forces when it comes to determining market prices. Fear drove several countries to put in place export bans to protect domestic food markets to the detriment of other food-importing countries. Meanwhile, speculative capital flooded into food commodities as other economic activities diminished (real estate markets easing, energy prices declining, etc.). Meanwhile, the price of inputs such as fuel and fertilizer skyrocketed – largely due to the dramatic increase in the price of oil – creating even more upward pressure on prices.

The urban poor are highly susceptible to food market instability, largely due to three factors. First, most of the urban poor (and the poor in general) spend a large proportion of their household income on food. The Asian Development Bank estimates that 60 per cent of all income of the poor in Asia is spent on food. In the city of Dar Es Salaam, the amount is 80–85 per cent. By comparison, in Canada and the United States, the proportion varies from 6 to 15 per cent. Therefore, it is clear that a small change in price – let alone the doubling of the price of basic foods such as rice and grains as was seen in 2008 – can lead to a significant increase in hunger.

The vulnerability of city dwellers is exacerbated by two key facts highlighted by the FAO. First, urban diets are more dependent on tradable commodities vs. traditional diets and, second, there is less land owned by households on which to grow their own food for sustenance. In cities, the high cost of real estate has pushed many of the poor and recent migrants to the margins. They are forced to settle on polluted lands where they sometimes (in fact, often) produce food with certain health risks.

Beyond the fundamental food security argument, two other elements may influence the future of policy on

*Email: mredwood@idrc.ca

INTERNATIONAL JOURNAL OF AGRICULTURAL SUSTAINABILITY 8 (1&2) 2010

PAGES 5–6, doi:10.3763/ijas.2009.c5009 © 2010 Earthscan. ISSN: 1473-5903 (print), 1747-762X (online). www.earthscan.co.uk/journals/ijas

urban agriculture. The first is that security concerns are closely entwined with access to food. Between January and May 2008, food riots were recorded in 37 different countries – Niger, Senegal, Egypt and Haiti to name but a few. The uncomfortable truth is that if people cannot access enough food we are dealing with a security crisis, not only a health crisis. Second, urban agriculture is a stimulus for political and social strength as it is increasingly around urban farming that many poor urban agriculturalists organize. Farmer organizations are often the way that those practising urban agriculture access political strength.

The experience of dozens of research projects on urban agriculture that have been conducted with support from the International Development Research Centre suggests four important responses. First, cities should protect productive agricultural land – this means encouraging more dense forms of development and preventing sprawl.

Second, land policy in and around cities needs to be designed in a way that accepts agriculture as a legitimate land use. No longer can cities render agriculture an illegal activity. If it is legal, it can be regulated and risks can be managed. This can be supported through the establishment of favourable tax incentives for private owners of idle land to make it productive. In other words, community associations of farmers are able to access vacant land to produce food for market.

Third, cities should design, promote and maintain city markets (Kampala and Rosario are good examples) whereby local produce can be sold. This is a key way not only to help sellers reach their consumers, but also to ensure that health standards are maintained, since a few markets are more easily managed than several hundred points of sale in an informal system of markets. An important research gap that needs to be addressed is the lack of comparative analyses done of the economic value of urban agriculture. Part of the problem is that often the activity resides in the informal sector where data collection is challenging.

Finally, the establishment of food policy councils can be a helpful method to coordinate municipal responses to urban food security. A food policy council in government acts as an 'anti-poverty cell' in a sense, and can ensure that land use planning is coordinated with community development and health authorities for the benefit of food production.

What is clear is that urban agriculture is here to stay. While no panacea for any global level of food insecurity, it provides millions with some secure access to food and reduces their exposure to volatile changes in price outside their control. Provided there is a reasonable level of protection of health – and clearly international policy makers such as the FAO and WHO as well as a growing number of national and municipal institutions are becoming more proactive – then it is an activity which should be encouraged. Cities cannot be segregated from agricultural policy development as they are largely the drivers of demand in the sector.

Sustainable urban agriculture: stocktake and opportunities

Leonie J. Pearson[1]*, Linda Pearson[2] and Craig J. Pearson[3]

[1] Swinburne University of Technology, GPO Box 218, Hawthorn, VIC 3122, Australia
[2] Faculty of Law, University of New South Wales, Sydney, NSW 2052, Australia
[3] Melbourne Sustainable Society Institute, University of Melbourne, VIC 3010, Australia

This paper reviews research on urban agriculture which relates to the three dimensions of sustainability: social, economic and environmental. We propose that urban agriculture has three elements: urban agriculture in isolation; its interface with the people and environment within which it is situated; and its contribution to the design of built form. Additionally, we consider its scale: micro, meso and macro. The analysis draws attention to legal, social and economic constraints and opportunities. It suggests that future priorities for research should be directed towards (i) strategically identifying principles of sustainable urban agriculture that help policy makers to design resilient cities, e.g. using flood-prone areas for food and employment, and (ii) operationally trialling innovative institutional mechanisms, e.g. differential land taxes to support sustainable urban agriculture or payments for environmental services provided by urban agriculture such as carbon sequestration.

Keywords: economic, institutions, legal, resilience, scale, social

Introduction

Urban agriculture (UA) is the producer, processor and market for food, plant- and animal- sourced pharmaceuticals, fibre and fuel on land and water dispersed throughout the urban and peri-urban areas, usually applying intensive production methods (expanded from Smit *et al.*, 1996). It encompasses greenhouse cropping and intensive animal industries. 'The lead feature of UA which distinguishes it from rural agriculture is its integration into the urban economic and ecological system' (Mougeot, 2008, p. 9). UA includes horticultural crops (fruits, flowers), but it does not usually include amenity or landscape horticulture, either at the home-garden or parkland scale. This convention is retained in this article although we note that amenity horticulture provides an opportunity for UA – simply by changing to species of plants that produce food. In our review we have drawn on research in amenity horticulture where we did not find any in UA.

Food production in cities has a long tradition in many countries and the UNDP (1996) has estimated that UA produces between 15 and 20 per cent of the world's food. Urban agriculture, although practised in both developed and developing economies, often serves different purposes, e.g. recreation in the former and food security in the latter. We draw most of our analysis from the literature of the developed economies.

For this paper, and indeed this Special Issue, we focus on the production end of urban agriculture. We disaggregate it into three scales: micro, meso and macro, and categorize ownership as private, corporate or public entities (Table 1).

The paper is purposeful in that we ask the question: what are the priorities for research and public policy to assist UA's contribution in creating sustainable, resilient cities? We define resilient cities in line with Holling's (1973) definition of resilience as the capacity of the city (built infrastructure, material flows, social functioning, etc.) to undergo change while still maintaining the same structure, functions and feedbacks, and therefore identity.

The focus on the UA–city interface is relatively unique because the literature on sustainable cities, and the need for transformational change to the ecology of western cities, usually ignores the opportunities for UA to contribute to urban sustainability. For example,

*Corresponding author. Email: Ljpearson@swin.edu.au

INTERNATIONAL JOURNAL OF AGRICULTURAL SUSTAINABILITY 8 (1&2) 2010

PAGES 7–19, doi:10.3763/ijas.2009.0468 © 2010 Earthscan. ISSN: 1473-5903 (print), 1747-762X (online). www.earthscan.co.uk/journals/ijas

Table 1 | Scale of urban agricultural production

Scale	Examples of scale	Broad ownership categories of UA land and produce
Micro	• Green roofs, walls, courtyards • Backyards • Street verges	• Private, corporate • Private • Public
Meso	• Community gardens • Individual collective gardens (allotments) • Urban parks	• Private, on public land • Private • Public
Macro	• Commercial-scale farms, e.g. turf, dairy, orchard, grazing (e.g. horses) • Nurseries • Greenhouses: floriculture and vegetables	• Private, corporate • Private, corporate • Private, corporate

Note: Ownership is categorized as private when owned by individuals with fully assigned property rights; corporate when owned by shareholders so that decision making is collective or may be assigned to corporate officers; and public when owned by government and managed for social outcomes, e.g. schools.

four studies of initiatives to improve the sustainability of US cities identify the importance of green space but not agriculture or food production (Saha and Paterson, 2008), land-use controls to reduce carbon emissions omit UA (Brown *et al.*, 2005) and actions to improve urban health do not mention the possible contribution of UA (Kjellstrom and Mercado, 2008).

The disconnection between UA and city ecology may be attributed equally to the focus of planners on the built environment, and the narrow focus of researchers into UA. In an excellent review, Mougeot (2008) characterizes the history of research in UA in three phases: first, geographical accounts; then isolated surveys and (in our words) 'success stories' or advocacy; then institutional projects led by multidisciplinary teams focused on the UA system rather than UA as part of a city system. However, it is self-evident that UA is part of the city ecosystem, with some statistics to indicate its importance: for

example, employing 200 million people, engaging up to 80 per cent of urban households in some less developed countries (Vietnam, Nicaragua, Ghana, Cuba), contributing about 15 per cent of state vegetable and fruit production in some developed economies (Australia), and increasing dietary diversity of those who participate in UA in two-thirds of countries analysed (Zezza and Tascottiu, 2008). Additionally, the intensity of UA ensures it achieves high productivity: in Sydney, Australia UA accounts for 1 per cent of the land area and produces $1 billion in agricultural produce, accounting for some 12 per cent of the state's agricultural production (Sinclair *et al.*, 2004). In this paper we seek to enumerate these and a much wider range of outputs (both positive and negative) from UA in the context of the UA–city relationship; individual papers in this volume add richness and specificity to this analysis.

Context

We propose three elements to the UA–city relationship: urban agriculture in isolation; its interface with the people and environment within which it is situated; and its contribution to the design and construction of the built form of cities.

There are many external pressures that will cause changes in UA and the design of cities. First, there is growing acceptance that the structure and function of cities must change rapidly to respond to various drivers, e.g. resource scarcity, population pressure (urbanization) and climate change (van Ginkel, 2008). This implies an opportunity for UA, as a component of cities, to affect this change through the UA–built environment interface. Second, as cities become very large, issues such as the increasingly complex and costly food transport chains (in financial, infrastructure and energy terms), and the negative effects of the built environment, e.g. heat islands, cause researchers and policy makers to review the outputs from UA and perhaps place more emphasis on their benefits, or modify UA to minimize dis-benefits that arise from it. Third, there is a growing call for changes in the practice of agriculture itself to create systems that are integrated and deal with the by-products from food transformation and consumption (e.g. Pearson, 2007).

For UA to address these opportunities, there is a need to have two elements (i) knowledge, and (ii) institutional structures, e.g. policies, laws and incentives. When reviewing these elements it became clear to us that the context for UA is different in developed from developing economies; for example, the primary purpose may be recreational or social in a developed, cash-based economy, but subsistence food production in a developing country. Further, city planning and legal jurisdictions are so different between countries, and especially between developed and developing countries, that global generalizations are unlikely to be helpful and are very possibly misleading. As a consequence, this article does not seek to compare and contrast UA in developed and developing countries and we draw most of our analysis from the literature of developed economies.

The aim of this paper is to identify what knowledge or research priorities are necessary to achieve sustainable urban agricultural systems that contribute to resilient cities. An overview of current knowledge, categorized into social, economic and environmental systems and their institutional environment (consistent with UA sustainability research, e.g. Ellis and Sumberg, 1998) is given in the next section. This leads to identification of knowledge gaps and priorities. The paper places emphasis on institutional arrangements, e.g. laws and incentives for creating sustainable systems.

Current knowledge of sustainable urban agriculture

This section summarizes the knowledge that is available about urban agriculture. There is no single approach to assessing the sustainability of UA; however, three approaches have informed our investigation: (i) the Food and Agricultural Organization's framework for the evaluation of sustainable land management to open-space UA (Drechsel *et al.*, 2008b); (ii) spatio-temporary dynamics of UA (Drechsel and Dongus, 2010); and (iii) a methodological assessment of UA (Nugent, 1999a). These approaches emphasize the elements of sustainability – environmental, social and economic – and the institutional environment to achieve sustainability. The current body of UA work can be broadly grouped into either case studies (e.g. Sawio, 1998; Larsen *et al.*, 2008) or descriptive accounts of concepts or theories associated with UA (e.g. McBey, 1985; van Veenhuizen and Danso, 2007). A sample of current knowledge and the contributions of articles in this special issue is given in Table 2.

Sustainability depends on the 'institutional environment' in which UA operates. The institutional environment includes social norms and rules as well as the formal laws and protocols for urban agriculture, summarized in Table 3. The categorization in Table 3 is adapted from Coggan and Whitten (2008) and includes economic and legal instruments for UA. Important for

any investigation is the need to recognize that regulations and incentives need to reflect differences in context, i.e. legal frameworks and social norms. For example, implementation of a Purchase of Development Rights scheme requires that a landowner own 'rights' (e.g. water rights, mineral rights) that can be transferred separately from the title of land, so that the landholder can sell the right to develop to government or others. Ownership of land in the US includes a right to develop (Daniels and Bowers, 1997); however, there is no independent 'right to develop' in many countries, e.g. Australia (McKenzie, 1997). The jurisdiction-specific challenges for institutional and regulatory design need to be accommodated in a way that may not apply for economic incentives.

All instruments require a clear understanding of who 'owns' the productive capacity of land for UA. In Table 1 we simplistically divide ownership into private, corporate or public (government) and suggest that public ownership is most important at the macro and meso scales, although when aggregated these may amount to large areas within cities. Whereas all three ownership categories in Table 1 can initiate actions to make UA sustainable, all require government support and/or implementation, as shown in Table 3.

Knowledge gaps in urban agriculture

Knowledge gaps in UA occur in two areas. First in the social, economic and environmental attributes of UA (Table 4). Second, in the institutions which govern UA, which we address in the following text.

Information and research into the social, economic and environmental goods and services produced by UA is weakest where it relates to the opportunities for UA to impact on urban form (planning, design and construction), followed closely by the interface between UA and the built environment (Table 4). Noticeably, there was a lack of discussion about the dynamic aspects of cities and UA, i.e. their transformational and resilient qualities, such as the role UA might (or might not) play in the transformation to low-carbon economies.

With respect to knowledge gaps in institutional governance, we have identified five points that relate to opportunities for innovation and scale: innovative institutional frameworks for success. Development of institutional frameworks, regulatory requirements and economic incentives has recently occurred for the conservation of biodiversity in urban and rural contexts (Kelly and Stoianoff, 2009). In comparison, there has

Table 2 | Key scientific knowledge surrounding the social, economic and environmental attributes of urban agriculture

Urban agricultural goods and services		Comments/references	Contributions in this journal issue
Social	Food security, access	'Urban penalty' in access to sufficient and healthy (quality) food for urban poor, thought to be addressed by micro UA (Islam, 2004; Zezza and Tascottiu, 2008). Access to fresh food rated as highest benefit from micro UA by English allotment holders (Perez-Vazquez *et al.*, 2005). Contribution of urban vs. peri-urban vs. rural food supply (Drechsel *et al.*, 2007). Several location-specific quantitative studies of contribution of urban vs. peri-urban vs. rural food supply (e.g. Drechsel *et al.*, 2007).	Eriksen-Hamel and Danso, Karanja *et al.*, Mason and Knowd
	Diet and health	Reviews of city health usually omit consideration of UA (e.g. Harpham, 2009). UA should be able to address specific deficiencies in micronutrients and vitamins (now recognized as more important than gross dietary energy or protein, e.g. Gibney and Strain, 1999; Allen, 2003) which is particularly constrained in poor societies (Seinfeld, 2003). See also Personal well-being.	Karanja *et al.*
	Personal well-being/psychological purpose, fitness	Agricultural activity may offset decline in physical activity with ageing after childhood, especially in women. Activity reduces health problems, e.g. obesity, type 2 diabetes, some cancers, clinical depression (Miles, 2007). Physical, intellectual and psychological benefits can occur from direct involvement in UA (micro–meso scale) related to issues of 'gardening' (McBey, 1985).	van Leeuwen *et al.*
	Sense of place	Farming protected in urban greenbelts of capital cities to preserve national rural rural (Canada, Australia: Pearson *et al.*, in review); environmental values associated with person's place-experience (e.g. Brandenburg and Carroll, 1995).	van Leeuwen *et al.*, Sumner *et al.*
	Aesthetics	Mixed appreciative responses: micro UA often viewed as nuisance, trivial, unsightly (Mbiba, 1994), abundant plants in offices as bizarre (Hansen and Machin, 2008), gardening as a 'harmonious human–nature interaction' (Brady, 2006), but meso UA, e.g. low-intensity farming, more aesthetically pleasing than intensive agriculture (Schupbeck *et al.*, 2008). Familiarity with agriculture 'should be included in all environmental education' (Van Bonsdorff, 2005).	Condon *et al.*, van Leeuwen *et al.*
	Social inter-actions/community building	Particularly in low income areas with disadvantaged groups; organization through UA (micro and meso level) leads to enhanced development; i.e. information flow, neighbourhood watch, community cohesion (Mudimu, 1996; Armstrong, 2000; Brown and Jameton, 2000).	van Leeuwen *et al.*, Sumner *et al.*
	Personal skills	Micro UA can develop horticultural and communication skills (Perez-Vazquez *et al.*, 2005)	Seymoar *et al.*
	Urban planning	Buffer zones, greenbelts, heritage values, social identity and economic infrastructure of UA influence urban planning as shown in Ile-de-France (Fleury, 2002).	Lydecker and Drechsel, Condon *et al.*, Knight and Riggs, Mason and Knowd, Merson *et al.*
Economic	Employment and income in production	Significant employment creation of the underemployed for both income (meso–macro level) and subsistence (micro level) livelihoods (van Veenhuizen and Danso, 2007). In developed economies, government estimates for Australian states are that UA contributes 12–18% of value of vegetable industry; Cuba 60% (Premat, 2005).	
	Gender equity	In developing countries, e.g. Zimbabwe, it is predominately women who undertake UA, ensuring they have greater input and self-determination of household resources (Mudimu, 1996).	Seymoar *et al.*
	Highest productive use of land	Vacant and degraded sites are utilized for UA, ensuring that land has the highest 'value' potential (Madaleno, 2000) and UA on vacant sites can lead to increased surrounding house prices (van Veenhuizen and Danso, 2007)	Condon *et al.*, Merson *et al.*
	Employment and income in UA value added (aggregate benefit to society)	Meso- and macro-scale UA has potential for commercial market value and significant employment opportunities, but little work has been done to quantify these aspects (Nugent, 1999b): some examples of benefits to society from West Africa (Drechsel *et al.*, 2007).	Karanja *et al.*, Seymoar *et al.*
	Diversified industry base in cities	Micro and meso level UA (e.g. Sawio, 1998; Larsen *et al.*, 2008) can have significant enterprise development and value-adding potential to UA production, e.g. retail (local food markets), marketing and supply chain opportunities, although no scientific study of impacts has been estimated (Nugent, 2003).	Merson *et al.*
	Energy transport ('food miles')	Food may account for 40% of all road freight; fossil fuel used in food transport 'in most cases exceeds the energy consumed in (production)' (UK: Jones, 2002). Current rise in popularity of 'local' food, markets.	Knight and Riggs

Table 2 | Continued

Urban agricultural goods and services		Comments/references	Contributions in this journal issue
Environmental	Waste recycling	Food waste contributes 30–40% of municipal solid wastes (UK, Korea: Forkes, 2007; Lee et al., 2007). Composting and recycling in agriculture, gardens reduces nutrient losses (e.g. Forkes, 2007), may lead to soil nutrient accumulation (Khai et al., 2007), and reduces greenhouse gas emissions (e.g. Tsai, 2008). May cause problems, e.g. ammonia generation, acidic leachate (Lee et al., 2007).	Lydecker and Drechsel, Eriksen-Hamel and Danso
	Urban heat and air quality	Urban heat island (UHI) and air pollution, e.g. ozone mitigated by vegetation. Micro (building/canopy) and meso (boundary layer) effects. UHI over vegetation/parks c. 4–5°C cooler than built environment (e.g. Taha, 1997; Wong and Yu, 2005; Strathopoulou and Cartalio, 2007). Benefit/cost of tree planting in Chicago 3:1 (McPherson et al., 1994). Mitigating UHI generates savings of $5–10 billion p.a. cooling electricity, 20% reduction in ozone (US: Meier, 1997).	
	Carbon sequestration	Vegetation (crop UA) sequesters carbon; other agriculture may be net emitters due to methane emissions from livestock and manure. Total agricultural emissions decline with urbanization (e.g. Lebel et al., 2007); sequestration by vegetation may be only 0.2% of city emissions (Manila: Lebel et al., 2007); air mixing disperses possible low-CO_2 islands associated with parks (and crop-based UA) (Wentz et al., 2002). Shade trees have approximately double the mitigative impact of reflective roofs (Akbari, 2002). Green roofs reduce building carbon emissions by 9% (Roehr and Laurenz, 2008).	Several at micro level: green walls (Dixon et al.), Eriksen-Hamel and Danso
	Wastewater recycling and health	Health risks of irrigated urban farming and options to address them (Cole et al., 2008; Drechsel et al., 2008a).	Lydecker and Drechsel, Eriksen-Hamel and Danso, Merson et al.
	Wastewater filtration	Vegetation filtering of nutrients and heavy metals from wastewater from industrial, farm or human sources (e.g. Verhoeven and Meuleman, 1999; Vymazal, 2005); less consideration of harvesting of vegetables for urban food or fibre.	Lydecker and Drechsel, Merson et al.
	Malaria	Green open spaces can be resting places for Anopheles but due to widespread use of polluted water, seldom breeding places (Afrane et al., 2004; Klinkenberg et al., 2005, 2008).	
	Noise	Abatement/absorption especially by trees (logarithmic reduction with height, width of vegetation); vegetation significant in attenuating city noise (Delhi: Kumar et al., 2008). UA causes noise pollution related to farm operations; substantial research on noise from machinery in workplaces but no publications found on reception/perception of agriculturally emitted noise in urban areas, although legal thresholds for urban noise are commonplace.	
	Odour	Vast literature on emissions, detection and control of odour, including tainting of products, e.g. milk. Urban impacts, e.g. legislated separation distances between livestock and human residences.	
	Light	UA may create islands or buffers with low reflected light (e.g. cropping) or light pollution (greenhouses). Unable to find refereed papers on light pollution from greenhouses despite this being an issue for planning and location of greenhouses in urban areas.	
	Pesticides	Pesticides and faecal coliform in urban-produced vegetables in Ghanaian cities exceed health thresholds (Amoah et al., 2006).	

been relatively little research and development of frameworks and incentives for sustainable UA. There is heavy debate regarding how effective these institutional frameworks are (Robinson, 2009), though this debate has also not spread to UA.

The institutional knowledge gaps that relate to scale include:

1. Issues of scale in UA are well recognised in the institutional environment. At the macro scale of production, the primary challenge is in preserving land for agricultural use in peri-urban areas while minimizing conflict between different land uses. There are two key points to note. In the absence of regulation or economic incentives, residential spread puts pressure on rural land for subdivision into smaller holdings, for rural residential use or for hobby farms. Even in cases where regulation is enacted it is not sufficient to protect land for UA without other supporting institutions, e.g. incentives, policy (Alterman, 1997). This fragmentation of land holdings both impacts on the capacity to support sustainable agriculture, and leads to an increase in the value of land used for non-agricultural uses, making farming uneconomic (Sinclair and Bunker, 2007). It is difficult for zoning and other planning controls to address this in the absence of economic incentives. For example, the NSW State Environmental Planning Policy (Rural Lands) 2008 was intended to encourage agricultural activity (e.g. by allowing subdivision for primary production into lots smaller than would otherwise be permissible provided there is no dwelling on the lot). However, it does not address how to identify the point below which agricultural activity is not sustainable or leaves insufficient separation from other uses, and, more importantly, it does not apply to any of the local government areas in the Sydney basin, where much of New South Wales' perishable vegetable production occurs (Sinclair and Bunker, 2007).

2. Managing conflict between inconsistent land uses is necessary at all scales but there is little systematic research into best scale-dependent practices. The impact of agricultural activities on neighbouring residential uses can generate claims of nuisance, such as interference caused by noise, smell, chemical sprays or dust; while non-farm activities can impact on agricultural uses, for example, dogs (Daniels and Bowers, 1997). Right to farm legislation is one response used in the USA, to give farmers a defence to civil claims for impacts on nearby land from noise, smell and other intrusions, as long as they are carrying out 'good management practices' (Daniels and Daniels, 2003). This is a limited response to the symptoms of conflict and does not address problems of planning and design of urban development that is sensitive to agricultural operations and appropriate modifications to farming practices on the edge (Sinclair and Bunker, 2007). An alternative approach, the provision of accessible mechanisms for resolution of disputes between landholders, has been attempted in Western Australia (Western Australia, 1995).

3. At the meso and micro scale UA is less regulated than macro UA and hence has its own issues. Agricultural activities incidental to a residential use in urban areas would generally be permissible subject to restrictions based on nuisance to neighbours. At the meso level economic incentives need to be carefully designed. For example, a differential tax or rating regime may tax commercial-scale agricultural land at a lower rate, leaving land used primarily for residential purposes and incidentally for agriculture to be taxed at a higher rate (Kelly and Stoianoff, 2009). At a community level governments can provide land for community gardens, which take one of two models: an allotment model in which gardeners have exclusive right to the use of an area of land, or shared gardens in which a garden is cultivated in common (Marrickville Council, 2007). Continued operation of a community garden may be vulnerable to changes in priorities for land use and perceptions of public liability associated with, for example, food poisoning from contaminated sites (Marrickville Council, 2007).

4. What are the best public policy instruments for each scale to optimize benefits? Broadly, there appears to be a link between scale of UA and type of instrument employed. Voluntary and information instruments are targeted towards micro and meso UA production. However, macro UA relies heavily on regulation for implementation. Many of the economic and social benefits of UA (from Table 3) arise from volunteerism at the smaller scales while regulatory, publicly owned activities may produce the greatest environmental benefits.

5. Lastly, there is a gap in institutional knowledge around how to 'scale-up' the findings from UA case studies. For example, UA has been designed into the roof infrastructure of Roppongi Hills, Tokyo, Japan; it includes a rice paddy and vegetable garden as well as aesthetic gardens. Its success has been measured in the reduction of the 'heat island' effect, workers' stress levels, etc. but there is a question of how realistic this 'case study' is for all buildings in a city. This has not been

Table 3 | **Current known institutional mechanisms and instruments for sustaining agriculture in urban areas**

Mechanism	Instruments	Examples
Regulations: required actions	Government zoning without compensation Zoning and planning controls, e.g. minimum subdivision sizes	Toronto greenbelt, to preserve agriculture (2005) New South Wales policies for identifying significant state agricultural land, and assessing land-use conflict (New South Wales, 2008)
	Government acquisition	Ottawa; land leased back to farmers (1965–1968)
	Support for efficient production, e.g. pesticide use, through regulations	Right-to-farm legislation in all US states except Iowa (Daniels and Daniels, 2003)
Economic incentives	Purchase of Development Rights (PDR)	Seattle USA, 24 states with PDR programs (Daniels and Daniels, 2003)
	Offset benefits: exchange for, e.g., increased building density elsewhere	USA: transfer of development rights (Daniels and Daniels, 2003)
	Differentiated agrarian taxes	NSW: 'farmland' usually rated at lower rate (Kelly and Stoianoff, 2009)
	Government payment for delivery of environmental goods and services (e.g. clean water)	USA: purchase of clean water by New York State through various catchment-wide approaches, e.g. planning, land acquisition (CGER, 2000)
	Rebates for farm inputs, e.g. diesel	Australia: income tax legislation (McKerchar and Coleman, 2003)
	Government grants for UA	USA: Community Food Security Competitive Grants Program (Feenstra, 2002)
	Subsidies to sustain production in difficult circumstances	Exceptional circumstances payments for broadacre agriculture, but not UA
Voluntary actions for enhanced security of UA	NGO-initiated land trust protected by legal covenant/perpetual conservation easement	Ontario Farmland Trust California, Vermont, Colorado (Daniels and Daniels, 2003)
	Government provision of land (and perhaps free services, e.g. water) for, e.g., community gardens	Australian city farms and community gardens: variety of forms/security of land tenure (Marrickville Council, 2007)
	Established industry structure for clear supply chain integration	Large-scale supply contracts by, e.g., supermarkets, not applying to UA
	Government maintenance of food-producing facility, e.g. fruit-bearing street trees	Canberra, since around 1935
Information, advice, support and moral suasion	Voluntary decision to use land for food production, e.g. household gardens	USDAs 'People's Garden' initiative to encourage and inform use of land for gardens Illawarra Biodiversity and Local Food Strategy for Climate Change
	Government and industry supported extension workers in rural and not UA	Extension officers (provided by government or NGOs) used in Harare (Drescher, 1996) and USA to improve UA yields
	Support for locally produced food, e.g. farmers markets	Friends of the Greenbelt (Toronto) Hawkesbury Harvest Farm Gate Trail (Sydney)

Table 4 | Subjective assessment of the knowledge gaps in sustainable urban agriculture as related to: (i) urban agriculture; (ii) interface of UA with the people and environment in which it sits; and (iii) UA's contribution to the design of built form

	Urban agriculture goods and services	Urban agriculture	UA interface with the people and environment within which it sits	UA contribution to built form
Social	Food security, access	2 (food security in developing countries); 1 but growing awareness in developed economies	1	1
	Diet and health	1	1	2
	Personal well-being/ psychological, purpose, fitness	1 Economic impacts, e.g. absenteeism reputed to be very large, but anecdotal	1	2
	Sense of place	2	1	2
	Aesthetics	1	1	1
	Social interactions/ community building	2	1 Unaddressed, whereas activity and community coherence seem increasingly important with ageing population, rising city crime	2
	Personal skills	1 Likely to be important for up-skilling in developing economies; more recreational in highly educated societies	n.a.	n.a.
Economic	Urban planning	1 Some awareness among planners but little literature/generalized lessons about benefits/dis-benefits of incorporation of UA into planning	2 Most of the work on conflict	1
	Employment and income in production	2	1	2
	Gender equity	2	2	1
	Economic value of land	3 highest value of agricultural or vacant land but same note as for urban planning	1	1 Little work done on the competing land uses, e.g. residential vs. UA vs. mining
	Employment and income in UA value added (aggregate benefit to society)	1 Most work is done at the micro scale and does not deal with the whole city	1	1
	Diversified industry base in cities	1	1	1
	Energy transport ('food miles')	2	1	2

Continued

Table 4 | **Continued**

	Urban agriculture goods and services	Urban agriculture	UA interface with the people and environment within which it sits	UA contribution to built form
Environmental	Waste recycling	1	1	1
	Urban heat	2	2	3
	Carbon sequestration	3 although research does not compare UA vs. green spaces, i.e. impact of harvesting	1	1
	Wastewater recycling and health	1	1	1
	Wastewater filtration	2	1	1
	Malaria	1	1	1
	Noise, odour	1 but substantial research in rural agriculture	1	1
	Light,	1	1	1
	Air quality	1	1	1
	Pesticides	1 as for noise etc.	1	1

Note: The ratings were provided by all three authors using a three-point scale, where 1 is little UA work undertaken and 3 represented a comprehensive research assessment of the issue from the UA perspective.

addressed within the literature. By contrast, there are participatory methods for scaling-up community development that is based on UA (Seymoar *et al.*, this volume).

Priorities for future research

Two directions for future research priorities emerge: strategic, to develop principles for implementing public policy for city design; and operational, to enhance UA's contribution to sustainable cities.

Strategic research may include:

1. Design and test policies and systems which will maintain UA (and its environmental benefits) as part of the urban system. This contrasts with the current situation where policies are primarily designed to manage conflict between UA and the built environment. These systems could be created through community consultation and trialled within local government areas, as with the introduction of on-farm nutrient management in the Netherlands (Neeteson *et al.*, 2002; Wiskerke *et al.*, 2003).

2. Research to assess the contribution of UA to large cities. Contributions which should be assessed include: (a) production of local food, as FAO and national statistics do not capture much of UA (e.g. UNDP, 1996); (b) employment in UA to reduce social inequity (Wilkinson and Pickett, 2009); and (c) the value of lower energy food chains. Other opportunities are: (d) contributions to adaptation to climate change, for example through sequestration of soil carbon; (e) shifting cultivation on underused (i.e. vacant) land between development opportunities and on flood-prone sites; and (f) using UA as a natural buffer against natural disasters, e.g. Hurricane Katrina.

While these strategic priorities encourage a move away from the current emphasis on case studies, we recognize that local developments or models are necessary to provide quantitative inputs for appropriate scaling-up to whole-city design. For example, taking the principle of the need to design the resilience of cities to pressures such as population and climate change by designing agriculture (in this case, urban agriculture) as semi-closed systems (Pearson, 2007), it is possible to design integrated green space/built environments that totally recycle wastewater, re-use the vast majority of solid wastes, improve aesthetics, reduce heat, and improve personal well-being, reduce absenteeism and improve community (workplace or neighbourhood) cohesion. Estate developers have designed such systems but their implementation is, to date, partial (e.g. the London Docklands) because of municipal conservatism and/or insufficient independent evidence of their economic or social benefits. This calls for the need for partnerships between planners and developers and researchers to quantify operational issues.

Operational research may be prioritized according to the mechanisms identified in Table 3:

1. *Regulatory.* Research priorities for regulations fall into two categories: macro-scale land-use management, and micro- and meso-scale enterprise regulations. Greenbelts, green wedges and protected agricultural land within urban conurbations (e.g. London, Toronto, and Ottawa; Portland and Melbourne; and Vancouver, respectively) provide macro-scale examples of regulatory intervention to support agriculture with minimal research as to its effectiveness. City parklands, often on a similar scale (e.g. Adelaide, Canberra) provide parallel regulated spaces with no agricultural content. A research priority would be to study the societal and economic benefits arising from different legal constraints on land-use within the recreational and commercial continuum from passive parkland to UA. At smaller scales, a priority is to review legal requirements for the practice of UA. While it is common to impose additional constraints oriented to urban residents (e.g. restricting the use of herbicides, buffers against light pollution from greenhouses) other regulations applied to UA, such as separation distances from intensive livestock, relate to much larger scale rural enterprises.

2. *Economic.* The embracing of biodiversity by local, city and national governments has given rise to a mix of economic incentives and regulations. A research priority is to explore a parallel set of incentives (or, in places, disincentives) for UA. For example, in addition to private benefits of profit and well-being, local governments may obtain financial benefits through carbon trading, offsetting and differential rates that encourage UA, such as the planting of trees in allotments and rear laneways. These mechanisms, e.g. payments for environmental goods and services, would increase the economic rent obtained from maintained greenspaces rather than the current

perception that they are vacant land awaiting building 'development' which (alone) generates economic return. At the micro level, mechanisms to respond to the Kyoto CDM would be augmented by economic returns from, for example, fruit produced for local restaurants. Other examples of economically oriented research in UA that would benefit effective land use within cities include analyses of economic costs and benefits associated with community (or commercial company) use of vacant land for UA during the development of new housing estates, and investigating the issue of separating ownership of UA land from UA produce, and how these two entities can provide multiple benefits to individuals and local communities. Green roofs and walls, increasingly encouraged by city planners, also lack underlying economic studies that quantify their benefits or provide a basis for differential land taxation.

3. *Voluntary or community engagement.* Peri-urban and rural communes and government-supported food production projects abound, whereas community-supported action research based on UA is rare or at least unpublished. Nonetheless, there are several opportunities (Table 3). Priority research would provide data on participants' well-being and other social issues (Table 4). For example, the making over of some gardens at Vancouver city hall to UA in 2009 provides a 'laboratory' to research community engagement, benefits and dis-benefits. Results would provide a basis for scaling-up to public policy (priority 1, above).

4. *Information.* While UA will be enhanced by information addressing technical issues and community organization for resource-poor city dwellers in developing countries (e.g. Seymoar *et al.*, this volume), in our view the information needs in developed cities relate to (i) supply chains, particularly niche or specialist marketing and local branding, and (ii) securing resources, e.g. land and water for urban gardens, from municipalities. These relate to scaling-up micro-scale UA for its long-term impact. Industry bodies, consultants and sometimes government departments provide this advice for rural farmers. Who will take responsibility for urban communities? (This relates also to economic structures to secure land and UA.)

In conclusion, in our view, research priorities for affluent societies are largely related to legal and economic issues although they need to be based on quantification of the environmental and social benefits generated by UA. The priorities include all scales, micro, meso and macro, but call for UA to be situated within the conceptual framework of the planned, built environment, rather than as UA has been viewed historically, as a discrete and often competing use of urban land. While most other articles in this special issue necessarily deal with UA case studies, the research priorities arising from this paper reflect the growing need by policy makers to deliver integrated urban policies.

References

Afrane, A. A., Klinkenberg, E., Drechsel, P., Owusu-Daaku, K., Garms, R., Kruppa, T., 2004, 'Does irrigated urban agriculture influence the transmission of malaria in the city of Kumasi, Ghana?', *Acta Tropica* 89 (2), 125–134 (special issue).

Akbari, H., 2002, 'Shade trees reduce building energy use and CO_2 emissions from power plants', *Environmental Pollution*, 116, S119–S126.

Allen, L. H., 2003, 'Interventions for micronutrient deficiency control in developing countries: past, present and future', *Journal of Nutrition* 133, 3875S–3878S.

Alterman, R., 1997, 'The challenge of farmland preservation. Lessons from a six-nation comparison', *Journal of the American Planning Association* 63, 220–243.

Amoah, P., Drechsel, P., Abaidoo, R. C., Ntow, W. J., 2006, 'Pesticide and pathogen contamination of vegetables in Ghana's urban markets', *Archives of Environmental Contamination and Toxicology* 50 (1), 1–6.

Armstrong, D., 2000, 'A survey of community gardens in upstate New York: implications for health promotion and community development', *Health and Place* 6.4, 319–327.

Brady, E., 2006, 'The aesthetics of agricultural landscapes and the relationship between humans and nature', *Ethics, Place & Environment* 9, 1–19.

Brandenburg, A., Carroll, M., 1995, 'Your place or mine? The effect of place creation on environmental values and landscape meanings', *Society and Natural Resources* 8, 381–398.

Brown, K. H., Jameton, A. L., 2000, 'Public health implications of urban agriculture', *Journal of Public Health Policy* 21 (1), 20–39.

Brown, M., Southworth, F., Stovall, T., 2005, *Towards a Climate-Friendly Built Environment*, Pew Center on Global Climate Change, Arlington, VA.

Coggan, A., Whitten, S., 2008, *Best Practice Mechanism Design and Implementation: Concepts and Case Studies for Biodiversity*, Final report for the Australian Government Department of Environment, Water, Heritage and the Arts, CSIRO Sustainable Ecosystems.

Cole, D., Lee-Smith, D., Nasinyama, G., 2008, *Healthy City Harvests: Generating Evidence to Guide Policy on Urban Agriculture*, CIP/Urban Harvest, Makerere University, Lima.

CGER (Commission on Geosciences, Environment and Resources), 2000, *Watershed Management for Potable Water Supply: Assessing the New York City Strategy*, National Academy Press, Washington, DC.

Daniels, T., Bowers, D., 1997, *Holding Our Ground: Protecting America's Farms and Farmland*, Island Press, Washington, DC.

Daniels, T., Daniels, K., 2003, *The Environmental Planning Handbook for Sustainable Communities and Regions*, Planners Press, Chicago, IL.

Drechsel, P., Graefe, S., Fink, M., 2007, 'Rural-urban food, nutrient and virtual water flows in selected West African cities', IWMI Research Report 115, International Water Management Institute, Colombo, Sri Lanka [available at http://www.iwmi.cgiar.org/Publications/IWMI_Research_Reports/PDF/pub115/RR115.pdf].

Drechsel, P., Keraita, B., Amoah, P., Abaidoo, R., Raschid, S. L., Bahri, A., 2008a, 'Reducing health risks from wastewater use in urban and peri-urban sub-Saharan Africa: applying the 2006 WHO guidelines', *Water Science & Technology* 57 (9), 1461–1466.

Drechsel, P., Cofie, O. O., Niang, S., 2008b, 'Sustainability and resilience of the urban agricultural phenomenon in Africa (Ch. 8)', in: D. Bossio, K. Geheeb (eds), *Conserving Land, Protecting Water*, CABI Publishing, Wallingford, UK, 120–128.

Drechsel, P., Dongus, S., 2010, 'Dynamics and sustainability of urban agriculture: examples from sub-Saharan Africa', *Sustainability Science Journal* 5 (1), 67–78.

Drescher, A. W., 1996, 'Management strategies in African homegardens and the need for new extension approaches', *Proceedings of the International Symposium on Food Security and Innovations – Successes and Lessons Learned*, Stuttgart, March.

Ellis, F., Sumberg, J., 1998, 'Food production, urban areas and policy responses', *World Development* 26 (2), 213–225.

Feenstra, G., 2002, 'Creating space for sustainable food systems: lessons from the field', *Agriculture and Human Values* 19 (2), 99–106.

Fleury, A., 2002, 'Agriculture as an urban infrastructure: a new social contract', in: C. A. Brebbier (ed.), *The Sustainable City II: Urban Regeneration and Sustainability. Proceedings of the International Conference on the Sustainable City, Spain*, WIT Press, Southampton, UK, 935–944.

Forkes, J., 2007, 'Nitrogen balance for the urban food metabolism of Toronto, Canada', *Resources, Conservation & Recycling* 52, 74–94.

Gibney, M. J., Strain, S., 1999, 'Food and nutrition for all', *Lancet* 354, SIV 38.

Hansen, A., Machin, D., 2008, 'Visually branding the environment: climate change as a marketing opportunity', *Discourse Studies* 10, 777–794.

Harpham, T., 2009, 'Urban health in developing countries. What do we know and where do we go?', *Health and Place* 15, 107–115.

Holling, C. S., 1973, 'Resilience and stability of ecological systems', *Annual Review of Ecological Systems* 4, 1–23.

Islam, K. M. S., 2004, 'Rooftop gardening as a strategy of urban agriculture for food security: the case of Dhaka city, Bangladesh', *Acta Horticulturae* 643.

Jones, A., 2002, 'An environmental assessment of food supply chains: a case study on dessert apples', *Environmental Management* 30, 560–576.

Kelly, A., Stoianoff, N., 2009, 'Biodiversity conservation, local government finance and differential rates: the good, the bad and the potentially attractive', *Environmental and Planning Law Journal* 26, 5–18.

Khai, N. M., Ha, P. Q., Oborn, I., 2007, 'Nutrient flows in small-scale peri-urban vegetable farming systems in Southeast Asia – a case study in Hanoi', *Agriculture, Ecosystems & Environment* 122, 192–202.

Kjellstrom, T., Mercado, S., 2008, 'Towards action on social determinants for health equity in urban settings', *Environment and Urbanization* 20, 551–574.

Klinkenberg, E., McCall, P. J., Hastings, I. M., Wilson, M. D., Amerasinghe, F. P., Donnelly, M. J., 2005, 'Malaria and irrigated crops, Accra, Ghana', *Emerging Infectious Diseases* 11 (8), 1290–1293.

Klinkenberg, E., McCall, P. J., Wilson, M. D., Amerasinghe, F. P., Donnelly, M. J., 2008, 'Impact of urban agriculture on malaria vectors in Accra, Ghana', *Malaria Journal* 7, 151.

Kumar, G. V. P., Deqangan, K. N., Sarkar, A., Kumari, A., Kar, B., 2008, 'Occupation noise in rice mills', *Noise and Health*, 10 (39), 55–67.

Larsen, K., Ryan, C., Abraham, A., 2008, *Sustainable and Secure Food Systems for Victoria: What Do We Know? What Do We Need to Know?*, Victorian Eco-Innovation Laboratory, University of Melbourne, Melbourne.

Lebel, L., Garden, P., Banaticla, M. R. N., Lasco, R. D., Contreras, A., Mitra, A. P., Sharma, C., Nguyen, H. T., Ooi, G. L., Sari, A., 2007, 'Integrating carbon management into development strategies of urbanizing regions in Asia', *Journal of Industrial Ecology* 11, 62–81.

Lee, S-H., Choi, K-I., Osako, M., Dong, J.-I., 2007, 'Evaluation of environmental burdens caused by changes of food waste management systems in Seoul, Korea', *Science of the Total Environment* 387, 42–53.

Lydecker, M., Drechsel, P., 2010, 'Urban agriculture and sanitation services in Accra, Ghana: the overlooked contribution', *International Journal of Agricultural Sustainability* 8 (1/2), 95–104.

Madaleno, I., 2000, 'Urban agriculture in Belem, Brazil', *Cities* 17 (1), 73–77.

Marrickville Council, 2007, *Community Gardens Policy Directions for Marrickville Council* [available at www.marrickville.nsw.gov. au].

Mbiba, B., 1994, 'Institutional responses to uncontrolled urban cultivation in Harare: prohibitive or accommodative?', *Environment and Urbanization* 6, 188–202.

McBey, M. A., 1985, 'The therapeutic aspects of gardens', *Journal of Advanced Nursing* 10, 591–595.

McKenzie, F., 1997, 'Growth management or encouragement? A critical review of land use policies affecting Australia's major exurban regions', *Urban Policy and Research* 15, 83–101.

McKerchar, M., Coleman, C., 2003, 'The Australian Income Tax system: has it helped or hindered primary producers address the issue of environmental sustainability?', *Journal of Australian Taxation* 6 (2), 201–223.

McPherson, E. G., Nowak, D. J., Rowntree, R. A., 1994, *Chicago's Urban Forest Ecosystem*, USDA Forest Service Northeastern Forest Experiment Station, Newtown Square, PA. Cited by Solecki, W. D., Rosenzweig, C., Pashall, L., Pope, G., Clark, M., Cosa, J., Wienke, M., 2005, 'Mitigation of the effect in urban New Jersey', *Environmental Hazards* 6, 39–49.

Meier, A., 1997, 'Editorial: special issue on urban heat islands and cool communities', *Energy & Buildings* 25, 95–97.

Miles, L., 2007, 'Physical activity and health', *Nutrition Bulletin* 32, 314–363.

Mougeot, L. J. A., 2008, *Urban Agriculture: Definition, Presence, Potentials and Risks* [available at www.trabajopopular.org.ar/material/Theme1.pdf].

Mudimu, G. D., 1996, 'Urban agricultural activities and women's strategies in sustaining family livelihoods in Harare, Zimbabwe', *Singapore Journal of Tropical Geography* 17 (2), 179–194.

Neeteson, J. J., Schroeder, J. J., Ten Berge, H. F. M., 2002, 'A multi-scale systems approach to nutrient management research in the Netherlands', *Netherlands Journal of Agricultural Research* 50, 141–151.

New South Wales, 2008, State Environmental Planning Policy (Rural Lands) [available at www.legislation.nsw.gov.au].

Nugent, R. A., 1999a, 'Measuring the sustainability of urban agriculture', in: M. Koc, R. MacRae, L. J. A. Mougeot, J. Welsh (eds), *For Hunger Proof Cities – Sustainable Urban Food Systems*, IDRC, Ottawa.

Nugent, R. A., 1999b, 'Is urban agriculture sustainable in Hartford, CT?, in: O. Furuseth, M. Lapping (eds), *Contested Countryside: The Rural–Urban Fringe in North America*, Ashgate, London.

Nugent, R., 2003, 'Economic impacts', in: SIDA (ed.), *Annotated Bibliography on Urban and Peri Urban Agriculture*, 130–169.

Pearson, C. J., 2007, 'Regenerative, semi-closed systems: a priority for twenty-first century agriculture', *BioScience* 57, 409–418.

Perez-Vazquez, A., Anderson, S., Rogers, A. W., 2005, 'Assessing benefits from allotments as a component of urban agriculture in England', in: L. Mougeut (ed.), *Agropolis: The Social Political and Environmental Dimensions of Urban Agriculture*, IRDC, London.

Premat, A., 2005, 'Moving between the plan and the ground: shifting perspectives on urban agriculture in Havana Bay, Cuba', in: L. Mougeut (ed.), *Agropolis: The Social Political and Environmental Dimensions of Urban Agriculture*, IRDC, London, 153–186.

Robinson, D., 2009, 'Strategic planning for biodiversity in New South Wales', *Environmental and Planning Law Journal* 26: 213–235.

Roehr, D., Laurenz, J., 2008, 'Luxury skins: environmental benefits of green envelopes in the city context', *Transactions on Ecology and the Environment* 1, 149–158.

Saha, D., Paterson, R. G., 2008, 'Local government efforts to promote the 'three Es' of sustainable development: survey in medium to large cities in the United States', *Journal of Planning Education and Research* 1, 21–37.

Sawio, C., 1998, *Managing Urban Agriculture in Dar Es Salaam. Cities Feeding People*, Report 20, IDRC, Ottawa, Canada [available at www.idrc.ca/in_focus_cities/ev-2486-201-1-DO_TOPIC.html].

Schupbeck, B., Zgraggen, K., Szenencsits, E., 2008, 'Incentives for low-input land-use types and their influence on the attractiveness of landscapes', *Journal of Environmental Management* 89, 222–233.

Seinfeld, H., 2003, 'Economic constraints on production and consumption of animal products', *Journal of Nutrition* 133, 4054S–4061S.

Seymoar *et al.*, this issue.

Sinclair, I., Bunker, R., 2007, 'Planning for rural landscapes', in: S. Thompson (ed.), *Planning Australia*, Cambridge University Press, Melbourne, 159–177.

Sinclair, I., Docking, A., Jarecki, S., Parker, F., Saville, L., 2004, *From the Outside Looking In: The Future of Sydney's Rural Land*, Funded by a University of Western Sydney Regional and Community Grant.

Smit, J., Ratta, A., Nasr, J., 1996, *Urban Agriculture: Food, Jobs and Sustainable Cities*, UNDP, New York.

Strathopoulou, M., Cartalio, C., 2007, 'Daytime urban heat islands from Landsat ETM and Corine land cover data: an application to major cities in Greece', *Solar Energy* 81, 358–368.

Taha, H., 1997, 'Urban climates and heat islands: albedo, evapotranspiration, and anthropogenic heat', *Energy & Buildings* 25, 99–103.

Tsai, W.-T., 2008, 'Management considerations and environmental benefit analysis for turning food garbage into agricultural resources', *Bioresource Technology* 99, 5309–5316.

UNDP, 1996, *Urban Agriculture: A Neglected Resource for Food, Jobs and Sustainable Cities*, UNDP, New York.

Van Bonsdorff, H., 2005, 'Agriculture, aesthetic appreciation and the worlds of nature', *Contemporary Aesthetics* 3, 1–14.

Van Ginkel, H., 2008, 'Urban Future', *Nature* 456, 32–33.

Van Veenhuizen, R., Danso, G., 2007, *Profitability and Sustainability of Urban and Periurban Agriculture*, FAO, Rome.

Verhoeven, J. T. A., Meuleman, A. F. M., 1999, 'Wetlands for wastewater treatment: opportunities and limitations', *Ecological Engineering* 12 (1–2), 5–12.

Vymazal, J., 2005, 'Constructed wetlands for waste water treatment', *Ecological Engineering* 25 (5), 475–477.

Wentz, E. A., Gober, P., Balling Jr., R. C., Day, T.A., 2002, 'Spatial patterns and determinants of winter atmospheric CO_2 concentrations in an urban environment', *Annals of the Association of American Geographers* 92, 1–15.

Western Australia, 1995, *Agricultural Practices (Disputes) Act 1995* [available at www.slp.wa.gov.au].

Wilkinson, R., Pickett, K., 2009, *The Spirit Level*, Allen Lane/ Penguin, London and New York.

Wiskerke, J. S. C., Bock, B. B., Stuiver, M., Renting, H., 2003, 'Environmental cooperatives as a new mode of rural governance', *Netherlands Journal of Agricultural Research* 51, 9–25.

Wong, N. H., Yu, C., 2005, 'Study of green areas and urban heat island in a tropical city', *Habitat International* 29, 547–558.

Zezza, A., Tascottiu, L., 2008, *Does Urban Agriculture Enhance Dietary Diversity? Empirical Evidence from a Sample of Developing Countries*, FAO RIGA [available at www.fao.org/es/ ESA/riga/pubs_en.htm].

The multifunctional use of urban greenspace

Eveline van Leeuwen[1]*, Peter Nijkamp[1] and Teresa de Noronha Vaz[2]

[1] Department of Spatial Economics, VU University, Amsterdam, The Netherlands
[2] CIEO – Faculty of Economics, Universidade do Algarve, Faro, Portugal

This paper calls attention to the critical role of greenspaces in cities, while it overviews the many functions they provide. From a theoretical perspective, the utility function of urban greenspaces concerns multiple dimensions. Temporal, spatial and social aspects clearly need a taxonomic approach, which is also described in this study. Thus, the prominent goal of this paper is to highlight the importance of the multifunctional use of urban green areas. It is next argued that multitasking strategies may enhance the use of farmland within urban areas and that this may turn out to be a win–win situation, provided urban planners are able to understand the different motivations of local communities. These basic issues are essential in identifying and mapping out attractive developments for modern urban greenspaces.

Keywords: multifunctionality; taxonomy; urban agriculture; urban green

Introduction

In the framework of the present paper, we refer to urban greenspace functions as productive green areas that are able to deliver useful products (wood, fruits, compost, energy, etc.) as a result of urban green maintenance or construction. The presence of these spaces can create and increase the economic value of an area, inter alia by providing new jobs. Green areas, water bodies, open space and attractive landscape architecture are all ingredients of an attractive urban setting. In particular, high-quality landscape types can lead to a considerable increase in real estate values, e.g. through hedonic prices. Thus, in most developed regions, in contrast to poorer areas of the world, green urban spaces are supposed to provide aesthetic environments for recreation and leisure, for which tourism and welfare are the drivers of landscape management. In both situations, whether in poor or richer environments, positive externalities may be obtained and they are generally associated with sustainable practices in landscape architecture. Urban green and vegetation are increasingly seen as necessary complements to the built environment in urban areas.

Urban green in history

Throughout human history, as exemplified in Table 1, urban greenspace has fulfilled many different functions, from production to leisure. The first way green urban spaces were intentionally promoted was by constructing urban gardens. According to several historic sources, the Hanging Gardens of Babylon were built by Nebuchadnezzar II around 600 BC (e.g. Tulleken, 1988).

When medieval households started to consume grain, very little was obtained from the rural hinterlands. Instead, the city population grew their own agricultural products, most of it within the walls of their villages (Jacobs, 1969). There, most of the activities related to innovative agriculture started.

Since the early 18th century, urban architecture has emphasized the use of gardening as an ornamental tool, in order to create pleasant and beautiful towns where decorative public gardens have encouraged citizens to be interested in learning about the world and many of its botanic species.

Later, and because of several food crises, another very interesting gardening facility was promoted in many northern European towns, where communal

*Corresponding author. Email: eleeuwen@feweb.vu.nl
INTERNATIONAL JOURNAL OF AGRICULTURAL SUSTAINABILITY 8 (1&2) 2010
PAGES 20–25, doi:10.3763/ijas.2009.0466 © 2010 Earthscan. ISSN: 1473-5903 (print), 1747-762X (online). www.earthscan.co.uk/journals/ijas

Table 1 | **Different uses for urban greenspaces through history**

Time period	Uses for urban green
600 BC	Private power and social status
1300 AD	Innovative agriculture
1700 AD	Collective gardens for knowledge
1900 AD	Food production
2000 AD	Leisure and recreation
2010 AD	Health and ecology

areas were allocated to the local population by the municipalities for families to cultivate urban vegetable gardens (allotments). Nowadays, the tradition of such gardens still exists, and in large cities such as New York, Berlin, Paris, London, Amsterdam or Moscow, as well as in many other smaller European towns, such spaces colour the most of the towns' hinterlands. In general, these individual plots are reduced in size (they may be often no more than 50m^2), but they can nevertheless provide families with fresh seasonal products, health and other educational or environmental benefits (Armstrong, 2000).

Today, urban greenspace is an indispensable element of urban quality of life. Green areas are environmental – and sometimes historico-ecological – assets of great importance for any city, as they may be considered ecological corridors. The importance of 'urban green' closely related to urban agriculture and gardening has been clearly recognized in urban architecture (e.g. MacHarg, 1971), as is witnessed, for instance, by Ebenezer Howard with his Garden Cities, Charles Fourier with his Phalansteries, and Ernest Calleback with his Ecotopia. An extensive overview of past and current practices can be found in Baycan-Levent et al. (2009).

The spatial dimension of useful urban green

The intensity of land use, and its corresponding prices, does not leave much land for extensive, space-consuming agricultural activities in urban areas. Still, some remains on account of its positive effects on health and relaxation. This is the case for activities such as commercial farmlands, in particular health farms, allotment gardens, community gardens, school gardens and city farms. It is noteworthy that farmland located in urban areas (e.g. urban farms or children's farms) is at present increasingly seen as an urban recreation area where citizens can walk, cycle, or enjoy nature.

In addition, urban greenspaces and urban vegetation moderate the impact of the negative consequences of human activities by, for example, absorbing pollutants and releasing oxygen. Furthermore, they maintain a certain degree of humidity in the atmosphere; they regulate rainfall, moderate changes in temperature and curb soil erosion, all contributing to a healthier urban climate for both humankind and nature. Urban green forms the basis for the conservation of fauna and flora in cities and it contributes to the maintenance of a healthy urban environment by providing clean air, water and soil, thus improving the urban climate. In particular in the future, when climate is likely to change in certain urban areas in the world, maintaining the balance of the city's natural urban environment will be of utmost importance; one example is the 'urban heat island effect'. Finally, urban greenspaces preserve the local natural and cultural heritage by providing habitats for a diversity of urban wildlife and conserve a diversity of urban ecosystems – imperative requirements within the climate change agenda and the Millennium goals. All such functions have a welfare-enhancing impact, particularly if they have an appropriate mix of functions (see http://stats.oecd.org/glossary/detail.asp?ID=1699).

The social dimension of useful urban green

From the perspective of local communities, the vulnerability of a global world adds uncertainty and lack of trust, particularly in the food sector, because much of the economic game is defined by unknown international decision makers. For several active economic and social players, in particular consumers, the notion of a 'borderless world' is uncomfortable and many communities are looking for solutions to gain some protection from the instability of global economic forces, for instance, by developing ecological resources to better explore the opportunities provided by nature. One such form is urban agriculture. This is generally practised for income-earning or food-producing activities, but may also be associated with recreation and landscape management from the perspective of hobbyism. Urban agriculture makes a positive contribution to food security, food safety and energy savings by shortening the circuits that distribute food products. Immediate advantages such as freshness of fruit and vegetables, better choice of high-quality meat products and simple processes for food traceability all mark a new trend in urban consumption and behaviour.

Urban agriculture in developed countries

Urban agriculture in developed countries comprises often small pieces of land tucked away in the corners of the city, owned by the local authority or railway companies. These plots of land are rented to residents so that they can grow flowers or vegetables, often on their own parcel. Community gardens are usually maintained by a group of persons (the community), just like school gardens, which have an extra-educational function besides providing agricultural products – sometimes organically raised. Frequently, production costs surpass the harvested value. Nevertheless, fresh food and vegetables from one's own garden are usually more appreciated than products from the supermarket. Examples of such horticultural activities may be found and studied in many developed countries.

This is clearly exemplified by The Netherlands. This country has around 250,000 community and allotment gardens, which account for around 4000ha of land. In Amsterdam, about 350ha of land is used for urban gardens (CBS, 2007). These gardens used to be places where fresh products were grown for the urban population. After WWII, they increasingly became ornamental gardens in which residents like to garden, but also to spend time relaxing. In this country almost 1000 farms exist. From the research study of Hassink *et al.* (2007), it appeared that, in 2008, average annual revenue from healthcare activities on a non-institutional health farm was about €73,000, which would amount to annual revenues of €72 million for the total Dutch non-institutional health-farming sector.

Another example comes from the United States. Armstrong (2000) found in her study of community gardens in New York that the social value of urban greenspace is not negligible. This town has a long history of using community gardens to improve psychological well-being and social relations, to facilitate healing, and to increase supplies of fresh foods. During and after both world wars, community gardens provided increased food supplies which required minimal transportation. During the Depression, city lands were made available to the unemployed and impoverished by the Work Projects Administration; nearly 5000 gardens on 700 acres were cultivated in New York City through this programme.

London offers another example. There are around 30,000 active allotment holders gardening on 831ha of land, of which 111ha are in inner London. Traditionally, allotment gardening has been a pastime for low-income or retired men. Furthermore, there are 77 community gardens in London which are located on housing estates, near railways, on temporary plots and in community centres. Community gardeners grow mainly flowers and ornamental plants, although there is also some cultivation of fruit and vegetables (Garnett, 1999).

Urban agriculture in developing countries

From a global point of view, green urban spaces materialize according their different contexts. For instance, in poor countries where the world's largest cities can be found, it is necessary to take advantage of every opportunity to supply nutritionally adequate and safe food. Here, urban agriculture is generally practised for food-producing activities that generate self-employment, direct revenues or savings, thus contributing to greater social stability. Communities of practice have been studied and described in the municipalities of Montevideo (Uruguay), Quito (Ecuador), Curaca (Brazil), Santiago de los Caballeros (Dominican Republic), Texcoco (Mexico), Bamako and Ouagadougou (West Africa), Nairobi (Kenya) and Tokyo's Nerima Ward (Japan) (FAO, 2000). It has been observed that one reason for the development of urban agriculture is its adaptability and mobility compared to rural agriculture. The expansion of cities helps to bring a wave of novel opportunities that encompasses urban, peri-urban and rural activities.

Also, from the perspective of local communities and circuits of proximity, the vulnerability of a globalized world adds uncertainty and lack of trust to individual decision making, because a great deal of the economic game is defined by unknown international decision makers. According to Ward *et al.* (2008), this has led to new rules in most of the agricultural policies and responsible institutions. A transition from a *productivist* towards a *post-productivist* model is emerging, independent of the dominance of long-established international trade rules. For many active economic and social players such as consumers, the notion of a borderless world is uncomfortable, and many communities are considering the possibility of gaining a certain degree of protection from the instability of global economic forces by further developing ways to better utilize the resources that can be provided. One such way is urban agriculture.

A taxonomic approach for urban green evaluation

Urban green is often at the centre of the debate about urban sustainability, as it is so essential for the urban quality of life. This discussion of the meaning of various types of urban green usually prompts serious questions concerning the valuation of urban space (for an extensive overview, see McConnell and Walls, 2005). In ecological economics literature, it is customary to make a distinction between the 'use value' and 'non-use value' of the environment. 'Use value' refers to the economic functions of space, e.g. for recreation, growing vegetables, wood, etc., while 'non-use value' refers to intangible functions of space, e.g. aesthetic pleasure, psychological well-being, social interaction, etc. It is noteworthy that urban greenspaces offer access to, and use of, a great variety of (mainly positive) ecological functions.

The various functions of urban greenspaces clearly show that greenspaces have a complex and multidimensional structure, and contain important values that contribute to the overall quality of urban life. A taxonomic approach for urban greenspaces has been presented by Baycan-Levent et al. (2009). In this taxonomy, the authors have defined a variety of urban greenspace values classified according to five categories: (a) ecological values: intrinsic natural value, genetic diversity value, life-support value; (b) economic values: market value; (c) social values: recreational value, aesthetic value, cultural symbolization value, historical value, character-building value, therapeutic value, social interaction value, substitution value; (d) planning values: instrumental/structural value, synergetic and competitive value; (e) multidimensional values: scientific value, policy value.

The authors have also developed an operational taxonomy for the evaluation of urban greenspaces in parallel with their taxonomy of urban greenspace values (see Table 2). This taxonomic framework offers a systematic assessment that demonstrates the multidimensional nature of urban green areas.

Most of the values attached to urban greenspaces are non-priced environmental benefits which include, for example, pleasant urban landscapes, peace and quiet, an improvement of the urban climate, and hence potential recreational opportunities, as well as better health for urban residents. It should also be recognized that qualitative valuations of greenspaces are difficult to integrate into conventional assessment procedures. For non-monetary valuation of urban greenspaces, suitable research methods for classification and evaluation include: geographical information system (GIS) methods, multi-criteria decision methods, meta-analysis and rough set analysis.

It is obvious that the complex and multidimensional structure of urban greenspaces makes the description or design of a single 'best' evaluation model for urban greenspaces difficult. Increasing complexity in urban greenspaces requires an evaluation on the basis of multiple decision criteria and multiple effects in an urban policy context. This multidimensional evaluation may comprise monetary and non-monetary valuation methods for both quantitative and qualitative information.

A prospective view for urban greenspaces

In summary, it should be noted that urban greenspaces provide a range of benefits in various forms and offer a variety of opportunities to people. They reinforce the identity of towns and cities, which can enhance their attractiveness for living, working, investment and tourism, and therefore these spaces can contribute positively to both the quality of life and the competitiveness of small cities. However, it is not surprising that the urban green policy has – in the light of urban sustainability policy – attracted much interest in recent years (for an overview, see Baycan-Levent et al., 2009).

It is, therefore, no wonder that in various towns and cities, new programmes based on ecological approaches have been developed for the protection and management of nature in urban greenspaces. Moreover, policy makers and planners have started to pay significantly more attention to initiatives designed to foster sustainable development and to improve the quality of life in urban areas by the clean-up and redevelopment of underutilized brownfield sites. Actually, there has been a growing recognition among urban community groups and environmental organizations that brownfields have enormous potential for 'greening' city environments, through the implementation of parks, playgrounds, greenways and other open spaces.

On the other hand, by 2015 around 26 cities in the world are expected to have a population of 10 million or more. At the present time, to feed a city of this size – for example Tokyo, São Paulo or Mexico City – at least 6000 tonnes of food must be imported each day.

Table 2 | **A typology of various approaches to the valuation of urban greenspaces**

Values of urban greenspaces	Values of urban greenspaces from an economic perspective	Valuation methods
1. Ecological values		
Intrinsic natural value	Existence value	*Monetary valuation*: cost–benefit analysis, travel cost method, replacement costs, tourism revenues, production function, contingent valuation
Genetic diversity value	Bequest value	
Life-support value	Indirect use value	*Non-monetary valuation*: species and ecosystem richness indices, genetic difference, genetic distance, phenotypic trait analysis, biodiversity index, keystone processes, health index, ecosystem resilience and stability analysis, hierarchical structure, population viability analysis, eco-regions or eco-zones
2. Economic values		
Market value	Direct/indirect use value	*Monetary valuation*: market analysis, production functions, financial analysis, economic cost–benefit analysis, travel cost methods, hedonic price method
3. Social values		
Recreational value	Direct use value	*Monetary valuation*: travel cost method, tourism revenues, contingent valuation
Aesthetic value	Existence value	
Cultural symbolization value	Existence value	
Historical value	Bequest value	
Character-building value	Indirect use value	
Therapeutic value	Indirect use value	
Social interaction value	Indirect use value	
Substitution value	Direct use value	
4. Planning values		
Instrumental/structural value	Indirect use value	*Monetary valuation*: cost–benefit analysis, contingent valuation, hedonic price method
Synergetic and competitive value	Existence value	*Non-monetary valuation*: geographical information systems (GIS) method, multi-criteria decision method
5. Multidimensional values		
Scientific value	Indirect use value	*Monetary valuation:* financial analysis, cost–benefit analysis, cost-effectiveness analysis, tourism revenues, taxes revenues
Policy value	Indirect use value/existence value	*Non-monetary valuation*: performance analysis, multi-criteria decision methods, meta-analysis, value transfer, rough set analysis, fuzzy set analysis, content analysis

Source: Baycan-Levent *et al.*, 2009.

Such large cities seem to be mushrooming and becoming a major problem for humankind. If urbanization is indeed out of control, then the emergence of a new generation of very large cities may undermine any progress towards sustainable development (PANOS, 1999). Urban land use needs to be flexible in order to meet the many socio-economic and sustainability objectives of our complex space-economy. Which options promise to cope with these challenges?

Multitasking to promote an enhanced use of farmland within urban areas may turn out to be a win–win situation, as explained by Deelstra *et al.* (2001). However, a major constraint is that urban planners around the world need to be able to understand the different motivations of local societies, adopting attractive land-use solutions designed to meet their individual needs. Such an effort is not easy because, firstly, of the difficulties in finding a consensus among the diverse policy aims; and, secondly, to the ever-present strong pressures on land use within and around cities, which causes increasingly high land prices. As such, and in order to accomplish that task (under conditions of city expansion with congestive demands), policy makers are continually searching for tools to integrate resource management and planning, using all the links between rural and urban areas, intensifying them if possible, and predicting citizens' needs in rural and urban areas.

The spectrum of activities based on urban agriculture to promote the green urban world is limitless and depends almost exclusively on the creative nature of the local population and its entrepreneurial capacities and leadership. However, significant added value is required if such activities are to be competitive with traditional urban industry. In Europe, there are many business combinations based on urban agriculture that provide different land-use functions. Examples of such multitasking for the green urban world include (e.g. www.greenexercise.org): agricultural and livestock farms for educational purposes or health care; food production combined with recreation and wastewater treatment; aquaculture with water storage and water sports; organic food and beverages and high quality standards in farm production in association with pro-active tourism; museum-oriented activities related to innovative or scientific processes used in food products of farm origin; urban forestry offering health and microclimate benefits; and energy extensive crops allied to both recreation and educational goals.

One can only speculate about the future of urban agriculture. Nowadays, urban land use already involves the regeneration of decayed industrial areas, embracing the concepts of modern fine arts to build green recreation places. Or, there is the possibility of cropping with no land by the application of hydroponics, or even the use of skyscraper terraces to construct neo-Babylonian suspended gardens and tropical mini-forests. Perhaps submerged gardens or undersea farming may one day result from romantic visions of mankind?

In conclusion, the post-productivist model opens perspectives to many new farm types and new urban garden forms in cities, which rapidly change their general brownish cartographic colour into an exciting mix with splashes of bluish-green. In other terms, it is amazing how – just when the rural world is becoming multitasking (Vaz and Nijkamp, 2009) because of the emergence of multifunctional agriculture – urban areas are beginning to identify the important role of agriculture in reshaping the landscape architecture of cities and to put into practice the many new concepts for business farms and green land-use forms. In this way, urban green areas may contribute to world-wide strategies to create sustainable cities.

References

Armstrong, D., 2000, 'A survey of community gardens in upstate New York: implications for health promotion and community development', *Health & Place* 6, 319–327.

Baycan-Levent, T., Vreeker, R., Nijkamp, P., 2009, 'A multi-criteria evaluation of greenspaces in European cities', *European Urban and Regional Studies* 16 (2), 219–239.

CBS (Centraal Bureau voor de Statistiek) [Dutch Statistics] (2007) Heerlen.

Deelstra, T., Boyd, D., van den Biggelaar, M., 2001, 'Multifunctional land use: an opportunity for promoting urban agriculture in Europe', *Urban Agriculture Magazine* 4, 33–35.

FAO, 2000, *Urban and Peri-urban Agriculture on the Policy Agenda: Virtual Conference and Information Market*, Joint venture of the FAO Interdepartmental Working Group (IDWG) [available at www.fao.org/urbanag/].

Garnett, T., 1999, 'Urban agriculture in London: rethinking our food economy', Paper presented at the international workshop 'Growing Cities Growing Food', Havana, Cuba, October 1999.

Hassink, J., Zwartbol, C., Agricola, H. J., Elings, M., Thissen, J. T. N. M., 2007, 'Current status and potential of care farms in the Netherlands', *NJAS Wageningen Journal of Life Sciences* 55 (1), 21–36.

Jacobs, J., 1969, *The Economy of Cities*, Random House, New York.

MacHarg, I. L., 1971, *Design with Nature*, Doubleday, New York.

McConnell, V., Walls, M., 2005, *The Value of Open Space*, Resources for the Future, Washington, DC.

PANOS, 1999, PANOS Briefing 34, June [available at www.panos.org/].

Tulleken, K., 1988, *The Age of God-Kings*, Time-Life Books, Amsterdam.

Vaz, M. T. N., Nijkamp, P., 2009, 'Multitasking in the rural world: technological change and sustainability', *International Journal of Agricultural Resources, Governance and Ecology* 8 (2), 111–129.

Ward, N., Jackson, P., Russell, P., Wilkinson, P., 2008, 'Productivism, Post-Productivism and European Agricultural Reform', *Sociologia Ruralis* 48 (2), 118–132.

Empowering residents and improving governance in low income communities through urban greening

Nola-Kate Seymoar[1]*, Elizabeth Ballantyne[1] and Craig J. Pearson[1,2]

[1] International Centre for Sustainable Cities (ICSC), 205–1525 West 8th Avenue, Vancouver, BC V6J 1T5, Canada
[2] Melbourne Sustainable Society Institute, University of Melbourne, VIC 3010, Australia

This paper describes the objectives, process and outcomes of three urban greening projects and how they contributed to sustainable development. The first two were conducted in low-income communities in Bangkok, Thailand, and in Badulla, Matale and Moratuwa, Sri Lanka. The lessons from these two projects were then incorporated in a post-tsunami project in Moratuwa and Matara, Sri Lanka. In addition to achieving urban greening objectives, the projects developed and validated a conceptual framework of sustainable community development (the Four-Directional Framework) using participatory action research. This framework facilitated rapid community learning and development objectives such as urban greening and consequential poverty reduction, empowerment of women, and improvements to the environment. Urban greening, incorporating urban agriculture, was an effective tool to improve the relationship between local authorities and the residents of marginalized and low-income communities and build a foundation for continuing sustainable development initiatives and city-to-city learning.

Keywords: community capacity building; ICSC; International Centre for Sustainable Cities; outcome mapping; participatory process; peer exchange; peer learning; sustainable community development; urban agriculture; urban greening

Introduction

It has been an axiom since the Earth Summit in Rio in 1992 that sustainable development is both a goal and a process (United Nations, 1992). The International Centre for Sustainable Cities (ICSC) used a multi-sectoral approach that links central government, local government and the host community with external funds and technical knowledge so that governments and poor communities are engaged in a participatory process of urban greening. Lessons learned in workshops and peer exchanges were used to encourage self-reflection and the transfer of learning from community to community. The three projects outlined here are not only instances of urban greening as a sustainable community development strategy; they also reflect the evolution of ICSC's process of working with communities and cities into a conceptual model for sustainable community development, the Four-Directional Framework, which describes and guided the interventions.

Urban greening as referred to in this paper includes two aspects: agriculture to produce food for consumption or sale, and environmental improvements. Typically, the food comes from home gardens and includes local vegetables and fruits, and plants for medicinal and cultural purposes. The environmental improvements that communities seek through urban greening include clean-up of waste, improvements in solid waste management including recycling, improvements to lanes and drainage, planting of trees for shade, and creating spaces for gathering. In the three projects, urban greening was intended as the core activity or catalyst for community development. The paper reflects on both the process and the environmental, social and economic impacts arising from these community-based greening initiatives.

Methodology

Each project was undertaken in a relatively short time frame (between 18 and 36 months). The three projects

*Corresponding author. Email: nkseymoar@icsc.ca

INTERNATIONAL JOURNAL OF AGRICULTURAL SUSTAINABILITY 8 (1&2) 2010

PAGES 26–39, doi:10.3763/ijas.2009.0467 © 2010 Earthscan. ISSN: 1473-5903 (print), 1747-762X (online). www.earthscan.co.uk/journals/ijas

shared features common among many international development organizations such as the partners who work with the Resource Centre on Urban Agriculture and Food Security (RUAF), including:

- local technical assistance
- formal relationship between ICSC, the local government and a local NGO
- formation of a multi-stakeholder project group
- gathering of baseline data using community residents
- establishment of a community green plan
- implementation of the plan
- evaluation of the results
- scaling out the results to other communities.

In the first project in Sri Lanka, Sevanatha (the local partner) used Community Action Planning (CAP) workshops to help participants assess their own needs. By the third project, Sevanatha had refined this strategy and used Community Livelihoods Action Planning, an adaptation of traditional CAP that focuses more on assets and livelihoods. Small grants were made available to seed community and livelihood activities, and small credit and savings groups were formed to sustain activities beyond the project timeframe. A major addition in the third project, which occurred after the tsunami, was the construction and management of Community Resource Centres.

Community selection

The selection of the communities involved stakeholders and partners in reviewing conditions of poverty, environmental challenges and the potential to benefit from the project, with a focus on the micro neighbourhood level rather than the larger urban context. The communities chosen were small (under 1000 residents), poor, and included informal settlements within the larger urban districts.

Green plan development and establishment of baseline data

Each community developed a 'green plan' that incorporated agriculture and composting to improve livelihoods and the local environment. Following community information meetings and the creation of community steering groups, the project coordinators worked with community residents to conduct an assets (and needs) assessment of the community, identify existing skills and interests, and establish a baseline of information (gender, age and number of people in households, level of income, nutritional data,

education). The approach combined aspects of various participatory field methods including CAP workshops.

Outcome mapping

Outcome mapping was introduced in the second project as an evaluation tool and incorporated in the third project. Mapping focused on changes in the behaviour, relationships, activities or actions of the people, groups and organizations directly involved rather than on traditional deliverables (Earl, 2001). The goal was to internalize monitoring and evaluation within the project and promote learning through self-assessment (see Fetterman, 2001). The method, developed by the International Development Research Centre (IDRC), has been formalized with a community of practitioners building case studies and sharing lessons learned.[1]

Gender focus

Although the first two projects targeted all members of the communities, it was observed that they were particularly successful in engaging and empowering women to take the initiative and responsibility in both urban greening initiatives and livelihood development, and significantly improved the dialogue between the local authorities (typically male) and local residents in the project (including many women). Although women from poor communities are largely excluded from leadership roles in both Thailand and Sri Lanka, food production is a culturally acceptable activity. The decision to focus on women in the third project in Sri Lanka resulted from these observations and the desire to bring the lessons from ICSC's post-earthquake project in Turkey (ICSC, 2000) into the post-tsunami situation in Sri Lanka.

Women's predominance in subsistence agriculture to meet home consumption needs and its implication for poverty reduction has been recognized for several decades (Boserup, 1970). In 1992 the Rio Earth Summit gave formal recognition to the role of women in a chapter of the Local Agenda 21 document, (United Nations, 2005) and in 1997 the UN Commission on the Status of Women made a series of recommendations on women including to 'involve women actively in environmental decision-making at all levels, including as managers, designers and planners, and as implementers and evaluators of environmental projects' (United Nations, 1997). Recognizing gender as a central issue in urban agriculture development, the website of the Resource Centre on Urban Agriculture and Food Security (RUAF) features references and discussion relating to the development of its

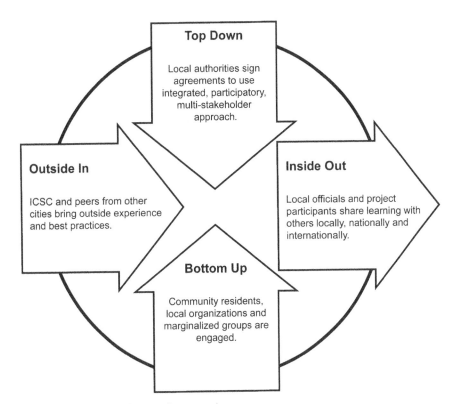

Figure 1 | ICSC's Four-Directional Participatory Framework

Four-Directional Framework

ICSC's Four-Directional Framework (Figure 1) outlines a process which: involves local governments and the business community (Top Down); mobilizes local residents and grassroots organizations (Bottom Up); introduces ideas and experiences from other communities (Outside In), and ensures ownership of the process and results by taking the emerging lessons and learning from the project to share with others locally, nationally or internationally (Inside Out).

The terms Top Down and Bottom Up have been common in the community development literature for decades. To the authors' knowledge the term Outside In was introduced by Taylor-Ide and Taylor (2002). ICSC's framework is similar to the Taylors' with the addition of Inside Out to describe its strategy for ensuring the ownership of the project and process by the participants by using a peer exchange (see below) approach to knowledge-sharing (Seymoar, 2004).

As applied in ICSC's work with cities, the Top Down step goes beyond including the municipality as an observer or participant, and entails a memorandum of agreement with the local council or authority to

own policy on gender issues in urban agriculture (see de Zeeuw and Wilbers, 2004).

formalize their commitment to the success of the initiative. Local officials are often not comfortable dealing with newly empowered communities and need help in doing so effectively. Bottom Up describes the engagement of grassroots participants in a participatory process which is asset based rather than problem focused. Described by Cooperrider *et al.* (2000), the Appreciative Inquiry approach focuses on strengths and possibilities rather than on problems, builds on what is working, and stresses the importance of asking the right questions. The Appreciative Inquiry approach is positive, essentially egalitarian and respectful of differences.

The Four-Directional Framework has evolved during ICSC's work with cities, regions and communities on a wide range of projects in the north and the south, including transportation, energy efficiency in buildings, heritage preservation, regional growth strategies, mining site restoration, waste management and urban greening. The framework has proven useful in these various contexts. There are many other established and emergent frameworks in specific arenas of development that are grounded in participatory process. As mentioned above, the work of RUAF and its partners recognizes the importance of the participatory approach in its Multi-Stakeholder Policy Formulation and Action

Planning (MPAP) process (Dubbeling and de Zeeuw, 2007), which is a valuable guide for engaging stakeholders who are represented by organized groups. ICSC's framework provides a guide for contexts where it is important to engage not only the formalized stakeholders such as NGOs and municipal authorities, but also marginalized residents whose voices may be hitherto unheard. Whereas the MPAP process usually proceeds from policy and regulations to action, the Four-Directional Framework describes situations where action in the community is encouraged simultaneously or even before consideration of policies and regulations, to build trust and produce immediate benefits.

Peer-to-peer learning

Peer exchanges are gatherings of people with a common purpose to share their knowledge and experience with others and to learn how others are approaching similar challenges. In the urban greening project in Bangkok, local residents took project learning to new communities. Later, Thai participants were invited to introduce ideas from their experience to Sri Lankan participants. Local and international peer exchanges and field trips were used in the Sri Lankan projects. In 2009 ICSC found that, when faced with a complex problem, 80 per cent of senior city administrators turn to their colleagues for help (Seymoar et al., 2009), further validating the observations made throughout the urban greening projects on the value of peer exchanges.

Project descriptions

Urban Greening Project, Bangkok, 2000–2001

This project brought together ICSC, the Bangkok Metropolitan Authority (BMA) and the Thailand Environmental Institute's (TEI) Grassroots Action Programme. It initially involved two low-income communities in central Bangkok: a Muslim community in Bangkok Noi *keht* (district) and a Buddhist community in Bangkapi *keht* (Figure 2). In the second phase, two more communities in Bangkapi *keht* and Pasri Charoen *keht* were added.

The specific objectives were to: educate residents about the benefits of urban greening; establish community groups and guide them in planning, establishing and maintaining urban green space; and develop and test a method of community involvement which supports the meeting of community needs while providing the environmental benefits of urban greening. The inclusion of marginalized groups was a central goal.

Meetings to engage members of the two communities were held in March and April 2000 to co-design the project (Fraser, 2002, p. 40). A group of 12 members from each community was selected to create a map of the potential green space, and an inventory and valuation of their local environment. The resulting Green Plans were implemented between July 2000 and May 2001 (Seymoar, 2003).

Urban Greening Partnership Program (UGPP), Badulla, Matale and Moratuwa, Sri Lanka, September 2003–September 2006

In 2003, ICSC and TEI, in partnership with the Sevanatha Urban Resource Centre, and the Clean Development Mechanism (CDM) and Faculty of Agriculture of the University of Peradeniya, undertook an urban greening programme in Sri Lanka modelled after the one in Thailand. Funded by CIDA, IDRC and the local Municipal Councils (MC), its purpose was to improve capacity, reduce poverty and address waste management issues using urban greening as a strategy.

In the first year of the programme, four neighbourhoods were selected in the three project cities, Badulla, Matale and Moratuwa[2] (Figure 3). The selected communities represented different agro-ecological, socio-economic and environmental conditions, and all suffered from poverty. The amount of green space in each of these neighbourhoods was on average less than 2.7 per cent, and they shared the following problems: insufficient waste and wastewater handling facilities, wastewater contamination, and dumping of solid waste in vacant lands, canals and streambeds. In the second year, the project was scaled out into six new communities.[3] In these 10 participating neighbourhoods in the three cities, approximately 3500 residents were impacted by the programme.

Meetings in September 2003 with the project partners, MC members and community residents gained community support. A door-to-door survey of approximately 250 households in each community was conducted to obtain a poverty profile.[4] During January and February 2004, a CAP workshop was held in each community to engage residents in discussions about how they would like to develop the lands made available by their respective MCs. Agriculturalists assisted participants in determining which plant species were best suited to their environments. A total of 163 residents participated, ranging from 49 to 63 per community.

Recognizing that the experience of the Bangkok communities could accelerate community learning in Sri Lanka through both technical advice and experience

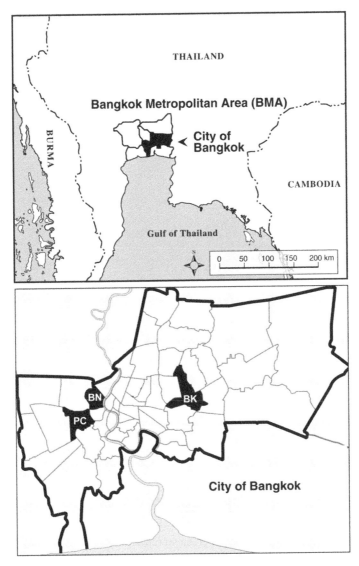

Figure 2 | (a) Location of Bangkok within Bangkok Metropolitan Area; (b) location of Bangkok Noi (BN), Bangkapi (BK) and Pasri Charoen (PC) *kehts* within Bangkok

in community organization, the project engaged TEI to provide its expertise as required on community greening and school training (ICSC, 2006a).

Each community developed its own Green Plan and decided what vegetables or plants it would grow. Chillies, for example, were chosen in Matale because they were selling for a good price, whereas Badulla chose shade trees. Based on the communities' needs, home garden plots as well as community gardens were created with seed and plant donations from the university. Planting occurred in April and May 2004, and community residents assumed responsibility for the maintenance of the trees, plants and flowers. Sevanatha

and university staff visited the sites on a regular basis to monitor the process and provide consultation. A Lessons Learned Workshop was held in August to reflect on the results to-date, discuss the next steps, and encourage other communities to initiate similar projects.

Outcome Mapping was brought in early in the second year to focus on the change sought in behaviours and relationships through the urban greening process (Table 1). The third year focused on analysing and disseminating results via the internet, factsheets and pamphlets. A regional seminar to disseminate the lessons learned from the UGPP project to urban local

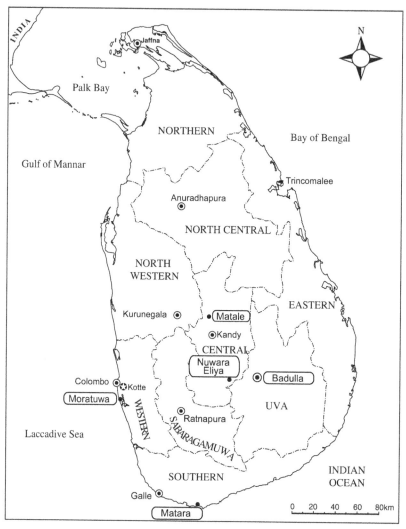

Figure 3 | Map of Sri Lanka indicating location of urban greening projects (b) and (c): Badulla, Matale, Moratuwa (UGGP); Moratuwa and Matara (CWRG)[5]

authorities, mayors and professionals entitled 'Urban Agriculture: Partnering for Poverty Alleviation' was held in April 2006 in Kandy, attended by 128 national and international participants. Following a final workshop in the spring of 2006, project staff and community leaders from Sri Lanka and Thailand presented their results at the World Urban Forum in Vancouver in June 2006.

Centring Women in Reconstruction and Governance (CWRG) in Moratuwa and Matara, Sri Lanka, June 2006–March 2009

When the tsunami hit Sri Lanka in December 2004, it wiped out the settlement in the Moratuwa community where the UGPP project had been located. No lives were lost but the residents were soon preoccupied with new priorities. In 2006, with individual donations and funding from CIDA, ICSC initiated a reconstruction project in Moratuwa and another affected coastal community, Matara, incorporating urban greening as a tool to build community capacity.

The purpose was threefold: (i) to improve women's participation in decision making and the local council's responsiveness to their needs; (ii) to equip women with income generation skills for livelihoods that enabled them to sustain their families and communities; and (iii) to improve the environment. A fourth element of the project involved building community resource centres in each community. Urban agriculture was seen as central to achieving objectives (ii) and (iii).

Table 1 | **Outcome mapping (expectation setting) by communities and municipal councils for projects in Badulla, Matale, Moratuwa (Sri Lanka)**

Communities: Low (L) = 0–35%; Medium (M) = 36–75%; High (H) = 76–100%

Expect to see	Like to see	Love to see
M Regular participation in meetings	H Allocating time for urban greening activities	L Exploring markets for their produce
H Adoption of proper waste management	H Harmony strengthened through formation of small working groups (Kalyanamitra Societies)	H Becoming an example for other communities
H More women than men participating	M Concern with health and nutrition values	L Being economically vibrant
H Move to home gardens	H Moving towards edible landscaping	
	L Moving towards commercial agriculture	
	M Providing inputs into policy formulation and decision making	
	M Improving creativity and adopting new technologies on their own	
	M Schoolchildren participating in urban greening activities	

Municipal councils: Low (L) = 0–35%; Medium (M) = 36–75%; High (H) = 76–100%

Expect to see	Like to see	Love to see
M Staff continuously learning about urban greening	H Earmarking physical, financial and human resources to urban agriculture	H Assisting in dissemination of concept of urban greening to other cities and other government institutions
M Encouraging officers connected to the programme	H Impressed with and approving of urban greening concept	M Consulting the community in policy making for urban agriculture
H Accepting the concept of urban agriculture as an alternative means of waste management	M Helping educate all sections of community on urban greening	
H Identifying with need for programme evaluation	M Officers providing extension services on urban agriculture and closely associating with the community	
	M Allocating finances from annual budget for urban agriculture	
	M Supporting and identifying with exemplary models of the project	
	L Help in seeking markets	
	M Understanding importance of identifying and obtaining assistance from institutions with the resources for urban agriculture	

Source: ICSC, 2006c.

ICSC brokered the funding and served as the project manager and Sevanatha again served as the delivery agent. The partners included GROOTS International (Grassroots Organizations Operating Together in Sisterhood) who brought expertise on mobilizing community women, and Builders Without Borders who advised on the construction of the resource centres. Advice on urban greening was provided through Sevanatha, TEI and the local department of agriculture, and through local peer exchanges with Matale, one of the Sri Lankan communities in the previous UGPP project.

Similar to the earlier projects, this one gathered baseline data through household surveys, formed small groups, and held workshops, training events and peer exchanges. A total of 61 activities were held over the project's term. Savings and credit groups were created that eventually became aligned with the Women's Bank. In each community, the small groups were the building blocks for the formation of a larger women's organization with an elected executive. These two women's organizations were formally registered with the local councils. The two community centres were designed using participatory design charrettes involving architects engaging with community members and other stakeholders.

Results

Urban Greening Project, Bangkok

Six progress indicators were used to evaluate the success of the project at the end of the second year: establishment of an urban green plan, community capacity building on environmental issues, poverty reduction, links with government, improvement in women's status, and development of a model for other communities. Successes were noted in all of these areas.

Both communities developed and implemented their Green Plans to create productive green spaces, including community gardens, shade trees and the planting of 20 new local species. In capacity building, Bangkok Noi and Bangkapi each developed a cohesive working group with 24 and 17 members, respectively, and more than 100 people contributing at various stages. Income generated by community gardens, as well as food for consumption, led to poverty reduction; for example, 10 Bangkapi gardens together on a monthly basis generated double the average monthly household income. Successful links with government were created, with the involvement of local government officials ensuring successful implementation – labourers were provided for three days, equipment was made available, a community development officer and a social welfare officer helped and attended workshops, and officials attended activities on six days.

The social outcomes included the engagement and empowerment of women. However, this was strongly dependent on local culture and religion. In the Bangkok Noi Muslim community, there was only one woman on the working group and women participated in only 30 per cent of activities, whereas in the Bangkapi Buddhist community, women were well represented on the community working group, a woman was treasurer of the project, and women and men were represented equally at all functions.

Based on the demonstrated success in these four communities, the BMA allocated funds to all 50 districts in the city for their own urban greening projects, and in 2003 integrated an urban greening strategy in the BMA's Community Education Planning Department and in the Department of Parks. Regarding the final indicator (model for other communities), an urban greening training manual was developed and distributed to 50 other communities, and a website provided access to the project detail. This institutionalizing of the project has been a significant factor in its sustainability in that the BMA and TEI continued to be engaged with the communities. In response to queries from ICSC in May 2009 TEI reported the following:

> The four communities in Bangkok still keep on working on environment issues ... Phoon Bumpen Community won the Green Community Award from BMA in the following year. Unfortunately, Building Together Community was not able to continue their urban agriculture along the canal as the land owner utilized that land for a condominium, although they still improved the small vacant land in their community for a recreation area. Ansorisunna Community also continued to pay attention to their small community garden and this area is still used as the nature learning area for school children. Lastly, Samakkee Pattana Community continues to look after the trees which they planted together in our project. So, now people in the community are having the big beautiful shading trees along the small road ... all communities state their satisfaction on the results especially on social and environment aspects (Kamuang, 2009).

Regarding the continued engagement of the BMA, TEI reported:

> BMA's Community Development Department was keen to apply the concept of the Urban Greening Project to all communities in 50 districts. So, they initiated the Green Community Award Campaign in 2005 to promote urban greening in communities ... In addition, the policy to make the city green has been issued by BMA continuously. According to

this policy some district offices have a budget to initiate the Urban Greening and Agriculture in their area and office (Kamuang, 2009).

Urban Greening Partnership Program (UGPP), Badulla, Matale and Moratuwa, Sri Lanka

The project (ICSC, 2006b,c) met its urban greening objectives in creating community gardens, residential green space, composting programmes and recycling programmes. Approximately 300 home gardens were created in the three initial communities, representing approximately five per cent of the communities' populations, with 100 using household waste composting. Three schools became committed to the project in one city and four schools in the other two. About 500 schoolchildren and several temple monks were engaged in urban greening.

In the first year, solid waste management was identified as a challenge. While many informal traders were picking through waste to collect plastics, glass and other discards for re-sale, there had been no attempt to organize recycling and make it a priority. Through the UGPP, the MCs provided the project communities with small separation centres. By the third year, most of the project communities had implemented composting programmes. Staff from the university provided training on 'how to compost', and Sevanatha sold composting bins to the communities with subsidies from local authorities.

The programme attracted large numbers of women, who represented 50–90 per cent of participants. At the well-attended urban greening exhibition in Badulla in July 2005, women's groups played the leading role in organizing the event, including food and activities. Exhibiting stalls were successful in selling their produce. Approximately 3000 students also attended.

The social benefits of the programme exceeded expectations. Outcomes included cohesion and empowerment, leading to the creation of women's groups, community micro-credit groups such as the Badulla Thrift and Credit Cooperative Society, and eco-learning centres at local schools. Local residents became urban greening champions and trainers for their communities. The formation of 21 Buddhist-based Kalyanamithra societies in the communities helped the participants to align their community urban greening efforts with their cultural aspirations.[6] Community capacity was strengthened further through the training of MC staff in each of the three initial communities, creating a pool of 30 individuals ready to support new urban

greening initiatives. Informal reports from municipal staff and residents indicated a number of social changes, including a shift from time spent unproductively or even in illegal activities to gardening, and the resolution of some long-standing feuds during the community efforts to build community gardens. A total of 22 people were trained in Outcome Mapping. The Badulla MC noted a reduction in waste collected due to the work of the MC Road Community in the collection and sale of recyclables.

By the close of the project, the Matale and Moratuwa MCs had created a separate budget line for urban greening in their 2006 budgets. The Moratuwa MC established a requirement that all newly constructed buildings incorporate green space into their designs. The Badulla MC agreed to provide a recyclable solid waste centre for the urban greening community (Ferguson, 2006). Mayors from the three project cities, Sevanatha and ICSC presented their knowledge and experience about urban greening, landscaping and solid waste management at a Mayors' Forum held in September 2006 with 11 of Sri Lanka's 18 MCs officially represented.

The project led to the establishment of lasting cooperative bonds among the municipalities based on commitment to sustainable community development: the five Sri Lankan cities joined ICSC's peer-learning-based Sustainable Cities: PLUS[7] Network and, six Sri Lanka cities later collaborated on a proposal for international aid to improve governance and peace at the local level.

Centring Women in Reconstruction and Governance (CWRG) in Moratuwa and Matara, Sri Lanka

Results were assessed in 2009 in three areas of sustainable development: improving environmental, social and economic well-being (ICSC, 2009).

Urban greening and improving the environment

Visible improvements to the environment consisted primarily of coastal clean-up and the clearing and improvement of lanes and drainage channels. Two hundred and nineteen residents (75 per cent women) cleaned drains and improved 800m of lanes. Matara hosted a beach clean-up campaign in which 40 women and 60 men participated. In poor communities prone to flooding, these basic infrastructure improvements have enhanced the quality of life of residents, and increased land values and the return in taxes to the MCs from a comparatively small investment.

Urban agriculture was stimulated, with 140 women receiving training in home gardening, composting and waste management. As a result, 60 women established home gardens with 20 composting their household waste. One of the women won first prize in a district gardening competition. The major agricultural enterprise was mushroom production; 55 women are earning incomes from this. A total of 30 women started making and using liquid organic fertilizer.

Improving women's participation in decision making and the local Municipal Council's responsiveness to their needs

Local women formed women's societies which were registered with the local MCs, the first women's organizations to be so recognized. In Moratuwa's Samarakoon Watte Ward, one of the city's poorest, the women's society Ran Arunalu had 112 members, 11 savings groups and 48 youth members, representing 36 per cent of the women in the ward. The Jayashakti Society in Matara's Kasiwattepura ward had 126 members (67 per cent of the women in the ward), eight savings groups and 52 youth members.

Moratuwa MC recommended that the Ran Arunalu society undertake the implementation of 25 housing projects on behalf of the Ministry of Urban Development and Sacred Area Development. Ran Arunalu managed the fund, monitored progress and completed the houses within the target dates. The Matara MC engaged members of the Jaya Shakthi society as trainers for the city-wide programme. Women have regularly attended meetings on environmental issues led by the Public Health Inspector in Matara, and started constructing waste separation bins for the MC. During the Lessons Learned workshop in December 2008, which included all partners and stakeholders, the participants concluded that 'women through their societies increased the visibility of the community issues with the local authority [MC] and were able to negotiate improvements, most notably regarding settlement upgrading, women's representation in the municipal committees, and access to livelihood opportunities and other social programs' (Tamaki *et al.*, 2009). The women's societies also provided advice based on their experience to other disaster-affected groups in Sri Lanka and India, and presented their experiences at international meetings in South Africa and the Philippines.

Although not measured in the project outcomes, major and significant results included a new pride of place and a changed sense of agency among the women and youth in cultures that are traditionally dominated by men. From hiding in doorways to making PowerPoint presentations at meetings with MCs and foreign delegations, community women emerged as leaders, assuming responsible positions in their own banking system, running community centres and sharing their lessons with other communities. The neighbourhoods are simply better and nicer places to live and this has been reflected in increased land values.

The Moratuwa Community Resource Centre (CRC) was opened in October 2008 and the Matara Centre was opened in February 2009. The CRCs are emerging as women-run and women-managed places for a variety of activities including health clinics, banking and livelihood training. In both centres, the women's involvement with home gardening led to their engagement in providing plants for the roof gardens. Although in Moratuwa there has been a reluctance on the part of the MC to sign a post-project formal agreement with the partners to ensure the continued running of the CRC as a women's centre, these two centres are cited as successful models of engagement and programming among other post-tsunami centres.

Equipping women with income generation skills for livelihoods that enable them to sustain their families and communities

Building on Badulla's experience in the UGPP program, one of the most significant decisions made early in the project by Sevanatha, in consultation with the community women and partners, was to establish savings and loans groups and to work closely with the Women's Bank to establish branches that could sustain the women's income generation activities beyond the CWRG project. As of March 2009 there were 80 Women's Bank members in Samarakmoon Watte with total assets of Rs556,160. They have dispensed 371 loans, of which 84 were for self-employment. There were 49 members in Kasiwattepura ward with assets of Rs147,000 and 49 loans, all for self-employment. An additional outcome has been the expansion of the Women's Bank into the neighbouring Willorawatte ward.

As later reported, 'participants felt that the livelihood development was the strongest and most immediate outcome, was practical, and is a strong foundation for further organizing and community development' (Tamaki *et al.*, 2009).

Although the project ended in March 2009, the activities in the two project areas are continuing with support from Sevanatha, the Women's Bank and the Community Livelihood Action Plan Network.

Discussion

The outcomes of these three projects fall into two categories: (1) the environmental, social and economic outcomes from the urban greening projects as catalysts for community capacity building and sustainable development; and (2) the systematic inclusion of peer learning in ICSC's model of development leading to the articulation and refinement of what is now referred to as the Four-Directional Framework.

The contribution of urban greening to sustainable urban development

The three projects demonstrate that urban greening can be a powerful tool of sustainable urban development. At the environmental level, interventions in the neighbourhoods improved drainage and restored natural spaces. At the economic level, gardens improved nutrition and, as micro-enterprises, improved livelihoods. At the social level, the urban greening projects empowered community residents, particularly women, to organize themselves and bring issues to their local MCs.

These benefits exemplify the multiple outcomes that have been noted from community development led through urban agriculture (e.g. Pearson *et al.*, this volume). Urban agriculture was effective in bringing local mayors and councillors into touch with their constituents through the largely non-controversial community green plans. Peer exchanges and the use of presentations to larger workshops and conferences involved community residents and elected officials travelling together. They saw innovations and came back with greater understanding of one another and more enthusiasm for trying out new ideas. These positive interactions built social capital and encouraged the MCs to respond positively to requests for assistance from the local residents. What is in place at the conclusion of the projects is a web of relationships among individuals and institutions that enables continued learning and cooperation. This trust may prove valuable in future times of stress.

Urban agriculture is increasingly accepted as a key strategy for poverty reduction (Mougeot, 2006), with recognition that participatory processes applied within a livelihood framework can lead to positive outcomes (Redwood, 2007). The importance of bringing stakeholders together, as well as the opportunity that urban agriculture provides for sustainable development, are also recognized by the RUAF Foundation (2009). While RUAF's process is organized into formal steps and focuses on policies, laws and regulations in the early stages, ICSC's process was more organic, beginning with urban agriculture or clean-up activities and moving to the policy level later. ICSC's results appear to be consistent with RUAF's.

In the projects described in this paper, the emphasis was on the lasting empowerment of the participants through urban greening – setting them up for continuing success. This empowerment is a crucial element that may not receive enough attention in the implementation of urban agriculture and other initiatives because it is the most elusive and difficult, involving working personally with the mosaic of individuals who make up a community and helping them coalesce into an effective working group. In all three projects, there have been benefits to the community that have been sustained over time. As noted, the Bangkok communities are still working on environment issues together eight years later due to the combination of a growing grassroots culture of environmental concern, support from the national environmental NGO and municipal funding. This continued institutional support can be attributed in part to the full engagement of these partners at every step of the process. In the Sri Lankan cities, sustainable community development and the support of urban greening have been embedded in the culture. Their membership in ICSC's Sustainable Cities PLUS Network not only symbolizes their commitment to join the vanguard of cities adopting sustainability practices, but also gives them access to an international community where they can seek peer support for their initiatives.

Gender and urban greening

The focus on women in the final project resulted in the construction of a permanent women's facility governed by local women in partnership with an established national NGO, aiming to provide a place and community anchor for learning and development initiatives to ensure that the investment in community capacity would be sustained and enhanced over time. While this was effective, it is understood that addressing the needs and potential of women separately may be an essential transitional strategy in Sri Lanka, but it is not sufficient in the long term. For this reason, this paper has not referenced the extensive recent work on mainstreaming gender in the specific urban agriculture agenda. However, as the work of Hovorka (2009) and others makes clear, gender dynamics are complex and often subtle, and achieving gender equality requires clarity and agreement on the definition, which is generally specific to the location or culture. In the long run, sustainable development, with food supply as a primary requirement, will require the education and empowerment of all residents as equally responsible participants.

Four-Directional Framework

The three projects provided a field laboratory for developing and testing the Four-Directional Framework. The projects started with an explicit agreement that they would be participatory, multi-stakeholder and holistic (involving all three elements of sustainable development: economic, social and environmental). From the beginning, the first Bangkok project was three-directional, comprising Top Down (local government), Outside In (ICSC and partners), and Bottom Up (community residents).

Although most development projects involving outside donors must usually ensure that the national government and local authority are aware of the initiative (Top Down), in these three projects care was taken to involve both elected officials and staff integrally in the projects. This led to municipal policy directions such as the Bangkok Municipal Authority's green city policy and their Green Community Award Campaign to promote urban greening. In the Sri Lanka UGPP project, the MCs made lands available for community agriculture; this did not take place in Bangkok's Building Together Community where the landowner later elected to build on the land he had earlier made available to the project. In Matara and Moratuwa, during the second Sri Lanka project, the municipality provided land for a community garden and a garbage sorting site. The initiative came from the grassroots community to obtain these benefits. In all cases, through dialogue and engagement with community members, the municipal authorities came to recognize the advantage of providing these benefits.

The participatory Four-Directional Framework can add to the impact of urban agriculture initiatives as a method that bridges the gap between local councils/municipal staff and community residents and enables communities to learn in a hands-on way from others. The most significant area for development projects to address is the empowerment of community residents to take charge of their own future, including food security. This is often neglected in the focus on technical improvements and the attention given to mature stakeholder groups that do not adequately represent the marginalized.

Peer-to-peer learning

The expansion of the project to other communities through a peer learning process brought into focus the potential power of this natural catalyst for change. Throughout all projects there were points of reflection triggered by peer learning exchanges (within, between and among communities), along with various Lessons Learned workshops where the emerging learning was tested and validated by the participants. This Inside Out element has emerged as an important formalized practice, in that every person who participated in the project is empowered and supported to become a teacher to others. Ownership of the results became widely held by group members. The potential of this approach to accelerate change is enormous, as it shifts the power to train and disseminate learning from a small number of experts to great numbers of grassroots residents who, with some facilitation and support, can share their learning broadly – locally, nationally and internationally.

Conclusions

The first of three lessons from these projects is that integrating urban greening initiatives with a community empowerment process is essential to bring about long-term benefits. Participatory approaches must go beyond basic engagement methods to ensure that the participants are active agents in their learning and take ownership of the outcomes. Collaborative goal setting for the projects is an important first step, but the process must be extended to include the articulation by the participants of ideas about how they want their community to be different or better at the conclusion of the project. Outcome Mapping is a useful and field-tested approach for conducting such a process, incorporating the Appreciative Inquiry method that builds on existing strengths and assets. Community empowerment involves learning not only at an individual personal level but also at a social level, where an empowering web of relationships among all the players is developed and cultivated. Modelling reflective community processes through events such as Lessons Learned workshops can become the nucleus for an adaptive community that takes on the challenge of creating its own prosperity.

A second lesson is that learning from peers from other communities accelerates the transfer of knowledge and skills. Trust is easier to build and maintain among equals – hence the power of peer-to-peer learning. These benefits were witnessed in the urban greening projects during the exchanges, and later when the local residents travelled to other cities and countries to tell their stories.

A third lesson is that anchoring the initiative within local institutions is central to ensuring that the work and the benefits can be sustained over time. Formally establishing women's organizations and having them legally recognized helps institutionalize the process. Local savings and loan societies supported by the local or national bank can serve a similar purpose, as can informal organizations such as the Buddhist-based Kalyanamithra societies in the Sri Lanka UGPP communities. Creating a lasting partnership between the community and a local NGO that can continue to provide collegial support and encouragement is another key ingredient, as is the support of the local municipality in the form of supportive policy and funding for urban greening initiatives. Finally, a physical centre can provide the focal point for ongoing community learning and a thriving local agriculture system.

Other challenges and areas for further exploration include the following:

- Capacity building takes time and an ongoing commitment to travelling together on this journey – a 5–10-year project time frame is more realistic to build a strong foundation and create lasting change.
- Financing at the micro level must be strategically designed to support individuals in their livelihood development, and a three-tier loans and grants scheme appears to be effective (savings and credit groups, a community revolving fund for infrastructure, and a fund to provide leadership training as well as seed grants for small community projects).
- Land tenure issues must be addressed early in the initiative to ensure long-term security of the sites for urban agriculture.
- Further research on the ways knowledge is transferred from community to community would be fruitful, exploring questions such as how much face-to-face interaction is needed to build trust and how this trust can then be supplemented by electronic and other tools.

Acknowledgements

We are grateful to the Thailand Environment Institute (TEI) in Bangkok and Sevanatha Urban Resource Centre in Sri Lanka, who contributed to the development of the Four-Directional Model, and to our partners in the execution of the three projects: TEI, the Liu Institute for Global Issues at the University of British Columbia, Sevanatha, the Clean Development Mechanism of the University of Peradeniya; GROOTS International, and Builders Without Borders; and CIDA, IDRC, the local MCs and other agencies and individuals for funding.

Notes

1. Outcome Mapping Learning Community [available at www.outcomemapping.ca/].
2. Badulla: Kanupallala community. Matale: MC Road community. Moratuwa: Jayagathapura and Siribaramenikepura communities.
3. Badulla: Badulupitiya community. Matale: Dola Road, Agalawatta, Vihara Mawatha and Kaludewala communities. Moratuwa: Sanwatsara Niwasa community.
4. Poverty profiles were found to be similar to those for Sri Lanka's largest city, Colombo (Jayaratne *et al.*, 2002).
5. The fifth city noted, Nuwara Eliya, was inspired by the example of the other cities to take on urban greening and other sustainability initiative. All five cities are now part of the ICSC: Sustainable Cities Network.
6. A Kalyanamithra is a person helping another with the genuine intention of seeking enlightenment through the work.
7. PLUS: Partners in Long-Term Urban Sustainability – a network of over 40 cities.

References

Boserup, E., 1970, *Women's Role in Economic Development*, Earthscan, London.

Cooperrider, D. L., Sorensen Jr., P. F., Whitney, D., Yaeger, T. F., (eds) 2000, *Appreciative Inquiry: Rethinking Human Organization Toward a Positive Theory of Change*, Stipes Publishing, Champaign, IL.

de Zeeuw, H., Wilbers, J., 2004, *Integration of Gender Issues in RUAF* [available at www.ruaf.org/node/888].

Dubbeling, M., de Zeeuw, H., 2007, *Multi-stakeholder Policy Formulation and Action Planning for Sustainable Urban Agriculture Development*, Working Paper Series #1, RUAF Foundation, Leusden, The Netherlands.

Earl, S., Carden, F., Smutylo, T., 2001, *Outcome Mapping: Building Learning and Reflection into Development Programs* [available at www.idrc.ca/en/ev-9330-201-1-DO_TOPIC.html].

Ferguson, J., 2006, Urban greening partnership project [Bangkok]: outcomes and lessons learned – final technical and financial report (unpublished), International Centre for Sustainable Cities, Vancouver.

Fetterman, D. M., 2001, *Foundations of Empowerment Evaluation*, Sage, Thousand Oaks, CA.

Fraser, E. D. G., 2002, 'Urban ecology in Bangkok, Thailand: community participation, urban agriculture and forestry', *Environment* 30, 37–49.

Hovorka, A., de Zeeuw, H., Njenga, M., (eds) 2009, *Women Feeding Cities: Mainstreaming Gender in Urban Agriculture and Food Security*, Practical Action Publishing, Rugby, UK.

ICSC (International Centre for Sustainable Cities), 2000, *From Disaster to Development: Women at the Epicentre* (Turkey) [available at http://sustainablecities.net/htmdocs/turkey_epicentre/epicentre.html].

ICSC (International Centre for Sustainable Cities), 2006a, *Urban Greening Partnership Program (UGPP). Sri Lanka: Final Report* [available at http://sustainablecities.net/docman-resources/doc_download/64-sri-lanka-ugpp-report].

ICSC (International Centre for Sustainable Cities), 2006b, Urban Greening Partnership Program (UGPP) [available at http://

sustainablecities.net/projects-overview/projects-past/urban-greening-partnership-program].

ICSC (International Centre for Sustainable Cities), 2006c, 'Urban greening partnership program, Sri Lanka. Year 3: final report (unpublished)', International Centre for Sustainable Cities, Vancouver.

ICSC (International Centre for Sustainable Cities), 2009, 'Centering women in reconstruction and governance: Sri Lanka 2006–2009', Final report (unpublished), International Centre for Sustainable Cities, Vancouver.

Jayaratne, K. A., Chularathna, H. M. U., Premakumara, D. G. J., 2002, *Poverty Profile: City of Columbo – Urban Poverty Reduction Through Community Empowerment*, Municipal Council, Colombo.

Kamuang, T., 2009, Personal email to N.-K., Seymoar, May.

Mougeot, L. J. A., 2006, *Growing Better Cities: Urban Agriculture for Sustainable Development*, IDRC, Ottawa.

Redwood, M., 2007, *Agriculture in Urban Planning: Generating Livelihoods and Food Security*, IDRC/Earthscan, Ottawa.

RUAF Foundation, 2009, *Why is Urban Agriculture Important?* [available at www.ruaf.org/node/513].

Seymoar, N.-K., 2003, 'Thailand urban greening project: final report (unpublished)', International Centre for Sustainable Cities, Vancouver.

Seymoar, N.-K., 2004, 'The four-directional framework of sustainable development (unpublished)', Presentation to Partnership Branch, CIDA.

Seymoar, N.-K., Mullard, Z., Winstanley, M., 2009, *City-to-City Learning*, International Centre for Sustainable Cities, Vancouver [available at http://sustainablecities.net/docman-resources/doc_download/119-city-to-city-learning].

Tamaki, M., Pathirana, V., 2009, 'Centering women in reconstruction and governance: lessons learned (unpublished)', Workshop, Moratuwa, Sri Lanka, International Centre for Sustainable Cities, Vancouver.

Taylor-Ide, D., Taylor, C., 2002, *Just and Lasting Change*, Johns Hopkins University Press, Baltimore, MD.

United Nations, 1992, Earth Summit, Global action for women towards sustainable and equitable development (Ch. 24), *Agenda 21* [available at www.un.org/esa/dsd/agenda21/res_agenda21_00.shtml].

United Nations, 1997, *Earth Summit + 5* [available at www.un.org/ecosocdev/geninfo/sustdev/womensus.htm].

United Nations, 2005, *Plan of Implementation of the World Summit on Sustainable Development* [available at www.un.org/esa/sustdev/documents/WSSD_POI_PD/English/POI Toc.htm].

Strengthening capacity for sustainable livelihoods and food security through urban agriculture among HIV and AIDS affected households in Nakuru, Kenya

N. Karanja[1], F. Yeudall[2]*, S. Mbugua[3], M. Njenga[1], G. Prain[4], D. C. Cole[5], A. L. Webb[6], D. Sellen[7], C. Gore[8] and J. M. Levy[9]

[1] Urban Harvest/CIP SSA, P.O. Box 25171-00630, Nairobi, Kenya
[2] Ryerson University, School of Nutrition and Centre for Studies in Food Security, 350 Victoria Street, Toronto, ON M5B 2K3, Canada
[3] Egerton University, Department of Human Nutrition, P.O. Box 536-20115, Egerton, Njoro, Kenya
[4] Urban Harvest/CIP, Apartado Postal 1558, Lima 12, Peru
[5] University of Toronto, Dalla Lana School of Public Health, Department of Public Health Sciences, Health Sciences Building, 155 College Street, Toronto, ON M5T 3M7, Canada
[6] Hubert Department of Global Health, Rollins School of Public Health, Emory university, Atlanta, GA
[7] Dalla Lana School of Public Health and Department of Anthropology, University of Toronto and Department of Nutrional Sciences, University of Toronto, Toronto, ON M5S 2S2, Canada
[8] Ryerson University, Department of Politics and Public Administration and Centre for Studies in Food Security, 350 Victoria Street, Toronto, ON M5B 2K3, Canada
[9] University of Toronto, Department of Anthropology, 19 Russell Street, Toronto, ON M5S 2S2, Canada

The promotion and support of urban agriculture (UA) has the potential to contribute to efforts to address pressing challenges of poverty, under nutrition and sustainability among vulnerable populations in the growing cities of sub-Saharan Africa (SSA). This may be especially relevant for HIV/AIDS-affected individuals in SSA whose agricultural livelihoods are severely disrupted by the devastating effects of the disease on physical productivity and nutritional well-being. This paper outlines the process involved in the conception, design and implementation of a project to strengthen technical, environmental, financial and social capacity for UA among HIV-affected households in Nakuru, Kenya. Key lessons learned are also discussed. The first has been the value of multi-stakeholder partnerships, representing a broad range of relevant experience, knowledge and perspectives in order to address the complex set of issues facing agriculture for social purposes in urban settings. A second is the key role of self-help group organizations, and the securing of institutional commitments to support farming by vulnerable persons affected by HIV-AIDS is also apparent. Finally, the usefulness of evaluative tools using mixed methods to monitor progress towards goals and identify supports and barriers to success are highlighted.

Keywords: agriculture; food security; HIV/AIDS; livelihoods; peri-urban; urban

Introduction

Rapid urbanization, unemployment and poverty have led to an increasing dependence by the urban poor on urban agriculture (UA) as a key livelihood strategy (Rakodi and Lloyd-Jones, 2002; Maxwell, 1995; Maxwell *et al.*, 1999; Cole *et al.*, 2008b; Prain *et al.*, forthcoming). Agricultural food production by the urban poor can enhance food security, provide additional income, and reduce vulnerability to economic shocks, environmental degradation and chronic instability in access to basic resources (Maxwell, 1995; Dennery, 1996; Cole *et al.*, 2008b; Prain *et al.*, forthcoming). In sub-Saharan Africa (SSA), it is projected that by 2015 half of the population will be living in urban centres and that poverty will move

*Corresponding author. Email: fyeudall@ryerson.ca

INTERNATIONAL JOURNAL OF AGRICULTURAL SUSTAINABILITY 8 (1&2) 2010
PAGES 40–53, doi:10.3763/ijas.2009.0481 © 2010 Earthscan. ISSN: 1473-5903 (print), 1747-762X (online). www.earthscan.co.uk/journals/ijas

increasingly from rural to urban areas (Cohen, 2004). These changes come at a time when the social dimension of agricultural production and sustainability is being re-emphasized and popular attention to food security has been heightened following the dramatic rise in world food prices in 2008 (Lyson, 2004; Swaans et al., 2006; Bawden, 2007; Pralle, 2008).

The capability of a household to produce, consume and sell food depends on the complement of 'assets' or forms of capital at its disposal (Bebbington, 1999; Prain et al., forthcoming). Such assets include access to land for food production, equipment and seeds to cultivate, human health and knowledge to enable people to tend and produce crops, and a supportive set of social relations. Households also need to be free from social, institutional, legal or political barriers to food production and marketing. Securing assets and achieving support from institutions is particularly challenging for poor urban households affected by HIV/AIDS (Loevinsohn and Gillespie, 2003; Swaans et al., 2006). Persons living with HIV/AIDS (PLWHA) in resource-limited settings often lack access to the foods required for optimal food and nutrition while on antiretroviral therapy (Castleman et al., 2003), yet often lack the energy to engage with institutions or obtain assets to produce such foods. Hence, potential negative relationships between HIV/AIDS and food and nutrition security are mediated through livelihoods (Gillespie and Kadiyala, 2005; Masariala, 2007). Participatory and interdisciplinary strategies to mitigate the impact of HIV/AIDS on livelihoods, food security and agricultural sustainability remain underdeveloped though promising (Swaans et al., 2006, 2009; Panagides et al., 2007; AED, 2008). Strategies must include interventions to reduce vulnerability to economic shocks, environmental degradation and stochasticity in resource access due to a range of insults that may originate at the global, national, regional, community or household level (Loevinsohn and Gillespie, 2003).

Here we describe the development and implementation of an international collaborative project to strengthen agricultural sustainability, social assets, food security and livelihoods among HIV/AIDS-affected households in the city of Nakuru, Kenya. The project came to be called SEHTUA for 'Sustainable Environments and Health Through Urban Agriculture'. We drew on documents and project notes, reports, monitoring and evaluation activities and meeting minutes to set out a timeline using RAPID methods (www.odi.org.uk/RAPID/), similar to that described in a case study in Kampala (Hooton et al., 2007). Figure 1 summarizes key external events, policy activities, research and

capacity building, partnerships and funding, since 2003. The accompanying narrative provides context and additional information.

Origins of SEHTUA

Urban Harvest

The Consultative Group on International Agricultural Research (CGIAR)'s system-wide programme on urban and peri-urban agriculture, Urban Harvest (UH), is hosted by the International Potato Centre (CIP). System-wide programmes seek to catalyse sharing of disciplinary skills in different international and national research organizations in collaborative efforts with other stakeholders. Urban Harvest is the only system-wide programme addressing the reduction of food insecurity and poverty in urban and peri-urban areas through more sustainable agriculture and improved natural resource management.[1] Urban Harvest's research for development strategy has been organizationally collaborative, interdisciplinarily constituted and action-research oriented. Such characteristics resonate strongly with recent thinking in urban governance and UA and is in keeping with the newer approaches to food security among those affected by HIV/AIDS (McCarney and Stren, 2003; Gillespie and Kadiyala, 2005; Prain, 2006; Swaans et al., 2009).

Project setting

Nakuru is Kenya's fourth largest municipality with a population of 302,784 (CBS et al., 2004). In line with earlier findings in six Kenyan towns (Lee-Smith et al., 1987), 35 per cent of Nakuru households farmed in town, 27 per cent grew crops and 20 per cent kept livestock (some doing both, hence sum >35 per cent) (Foeken and Owuor, 2000; Foeken, 2006). Common crops in Nakuru include maize, kale (sukuma wiki), beans, onions, spinach, tomatoes and Irish potatoes, while chicken, cattle, goats, ducks and sheep are common livestock (Foeken, 2006). Approximately 40 per cent of Nakuru residents are affected by poverty which limits their capacity to engage in UA (Kiarie, 2009). Poorer segments of the urban population (who have less access to land) are often less well represented among urban farmers than those who are better off, a trend particularly true for livestock keepers (Tevera, 1996; Mukisira, 2005). Indeed, a recent survey of mixed crop-livestock farmers observed a much higher rate of home ownership supporting a relationship between wealth and livestock farming in Nakuru (Karanja et al., forthcoming).

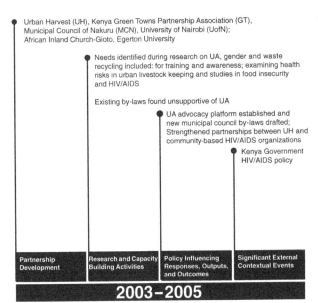

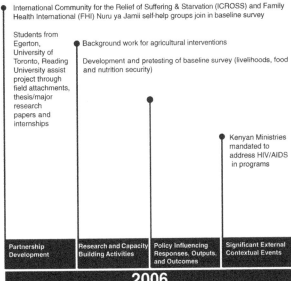

2003–2005

Research Funding
Community-based Research and development Centre on UA and Waste Management, Nakuru: IDRC 2003–2004
Local Participatory Research & Development on UA & Livestock Keeping in Nakuru, DFID 2004–2006

2006

Research Funding
Combating HIV/AIDS in urban communities through food and nutrition security: the role of women led micro-livestock enterprises and horticultural production in Nakuru town
CIDA-CGIAR Linkage Fund 2006–2009

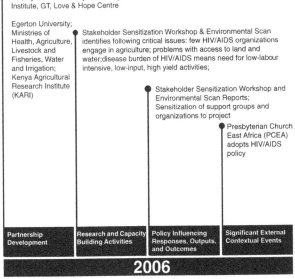

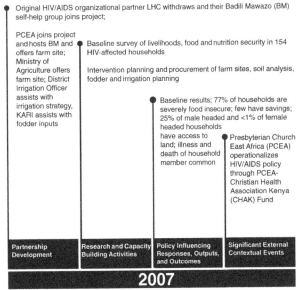

2006

Research Funding
Combating HIV/AIDS in urban communities through food and nutrition security: the role of women led micro-livestock enterprises and horticultural production in Nakuru town
CIDA-CGIAR Linkage Fund 2006–2009

2007

Research Funding
Combating HIV/AIDS in urban communities through food and nutrition security: the role of women led micro-livestock enterprises and horticultural production in Nakuru town
CIDA-CGIAR Linkage Fund 2006–2009

Figure 1 | Process of development and implementation of the Sustainable Environments, Health and Urban Agriculture Project (SEHTUA)

Early Urban Harvest collaborations

In 2004, UH partnered with the Municipal Council of Nakuru (MCN) and affiliated local groups to conduct a series of studies aimed at sustainable integration of urban solid waste with UA systems. *Community Based Research and Development Centre on Urban Agriculture and Waste Management in Nakuru* was carried out by Kenya Green Towns Partnership Association (Green Towns) and UH to address the issue of waste recovery and recycling for income,

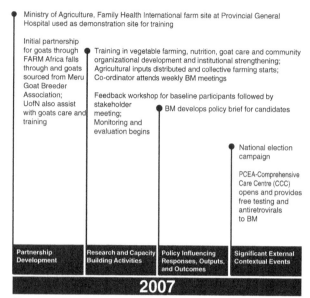

2007

Research Funding

Combating HIV/AIDS in urban communities through food and nutrition security: the role of women led micro-livestock enterprises and horticultural production in Nakuru town CIDA-CGIAR Linkage Fund 2006–2009

In-kind support from UofT based computing and software analysis facilities from Canada Research Chairs program, Ontario Innovation Trust and Canadian Foundation for Innovation

2008

Research Funding

Combating HIV/AIDS in urban communities through food and nutrition security: the role of women led micro-livestock enterprises and horticultural production in Nakuru town CIDA-CGIAR Linkage Fund 2006–2009

Two Canadian post doctoral fellows partially fund work from Canadian Institutes for Health Research fellowship grants

LHC rejoins project and works with BM to develop ideas for proposal submissions on UA

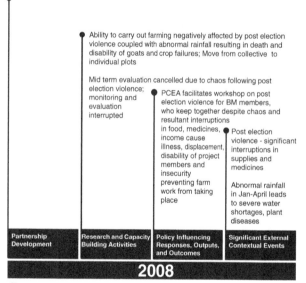

2008

Research Funding

Combating HIV/AIDS in urban communities through food and nutrition security: the role of women led micro-livestock enterprises and horticultural production in Nakuru town CIDA-CGIAR Linkage Fund 2006–2009

BM receives grant from Scotland Faith Based Organization (FBO) to fund development of education and support centre (BMGCHI); BM purchases land and begins construction of centre

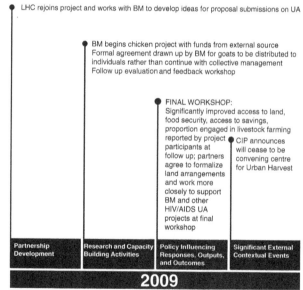

2009

Research Funding

Combating HIV/AIDS in urban communities through food and nutrition security: the role of women led micro-livestock enterprises and horticultural production in Nakuru town CIDA-CGIAR Linkage Fund 2006–2009

BM receives grant from Canadian FBO to fund chicken project at BMGCHI site

Figure 1 | *Continued*

employment, food security and environmental quality. This project involved a preliminary health hazard assessment study, conducted by the Department of Land Resource Management and Agricultural Technology, University of Nairobi. Meetings were held with active waste recycling groups, faith-based rganizations, NGOs, development partners, government departments, the MCN, and individual urban farmers. During stakeholder meetings involving the MCN, the Director of the Environment indicated that there was no support for UA in Nakuru's current environmental by-laws.

Urban farmers selected to participate in project training courses identified the need for research on urban organic waste, including livestock manure recycling and re-use. Hence, *Local Participatory Research and Development on Urban Agriculture and Livestock Keeping in Nakuru* was developed by the same group of partners to assist urban farmers and youth groups involved in waste recycling to improve their livelihoods and contribute to urban food security. Drawing on experience in another project in Nairobi, UH was able to share expertise in organic waste management and assist the process of reviewing municipal environmental by-laws (Njenga *et al.*, forthcoming).

Focus on HIV/AIDS-affected families

The Rift Valley provincial HIV prevalence stands at 7 per cent, higher than the national adult prevalence rate of 5.1 per cent. Women are disproportionately affected: 8.7 vs. 4.6 per cent among men nationally (NASCOP, 2008). In keeping with the Kenya National HIV/AIDS Strategic Plan (NASCOP, 2003), efforts to mitigate the impact of the pandemic on vulnerable households through an agricultural and nutrition for health project were deemed highly desirable. Building on UA–health linkages in the SSA region work in Kampala in particular (Cole *et al.*, 2008b), co-hosting an IDRC sponsored regional workshop on UA and health (Boischio *et al.*, 2006) and ongoing collaborative work with the University of Nairobi, UH and partners obtained funding in early 2006 from the Canadian International Development Agency CGIAR-CANADA Linkage Fund. Additional funding for SEHTUA was gleaned through research support to post-doctoral students from the Canadian Institutes for Health Research and the Canada Research Chairs programme.

SEHTUA aim and objectives

The aim of SEHTUA was to strengthen understanding of the links between agricultural sustainability and HIV/AIDS through an investigation of the potential of UA to mitigate the negative livelihood and food security effects of HIV/AIDS on households. Given the multidimensional, multilevel and multisectoral nature of the impact of HIV/AIDS on individuals, households and communities, SEHTUA

adopted an integrated approach with the following objectives:

1. Determine the role of crop and livestock production in urban livelihoods of HIV/AIDS-affected households;
2. Assess pathways by which crop and livestock production impact on food and nutrition security of HIV/AIDS-affected households;
3. Develop and evaluate strategies to improve livelihoods and food and nutrition security of HIV/AIDS-affected households, including through small livestock and horticultural production systems and dietary diversification and modification activities;
4. Enhance the capacity of local authorities and caregivers in food and nutrition security approaches in relation to HIV/AIDS-affected communities; and
5. Identify policy constraints and needs for strengthening livelihoods, food and nutrition security and social inclusion of HIV/AIDS-affected households.

SEHTUA partners and organization

Sensitization workshop

In keeping with the participatory nature of the project, the first major milestone was a sensitization workshop for stakeholders in mid-2006. The diverse set of stakeholders (see Table 1) included households, community organizations, community leaders, university researchers, international organizations, and local, provincial and national government officials. Working group discussions included (i) suitable agricultural technologies and interventions; (ii) potential stakeholders and mechanisms for involvement; and (iii) food and nutrition security issues of households with young children. Participants identified several challenges to practising urban agriculture for PLWHA and discussed potential mechanisms for overcoming them (SEHTUA, 2006).

Research institute, academic and agency partners

Urban Harvest provided overall project leadership. The Nairobi-based International Livestock Research Institute (ILRI) backstopped livestock studies. The Toronto-based Canadian universities Ryerson University and University of Toronto were responsible for leading livelihood, food and nutrition security studies. MCN oversaw HIV/AIDS national policy implementation within the district, Love and Hope Centre (LHC), a faith-based organizational partner working with PLWHA, provided contact with HIV/AIDS-affected persons

Table 1 | **Stakeholder organizations attending sensitization workshop by sector**

Sector	Organization
Self-help groups	Jamii Orphan Group Semeria Self-Help Group Together Hands Craft Self-Help Group Upendo Mpya Self-Help Group
Community-based organizations	AIC Rehabilitation Centre Catholic Diocese of Nakuru Kenya Green Towns Partnership Association Netreach Tumaini na Fadhili
Government	Kenya Agricultural Research Institute (KARI) Njoro Ministry of Agriculture Ministry of Health and Social Services Ministry of Livestock & Fisheries Development Ministry of Water and Irrigation Nakuru Municipal Council – Environment Nakuru Municipal Council – Public Health
Academic and research organizations	CIP Urban Harvest Egerton University International Livestock Research Institute (ILRI) Ryerson University University of Nairobi University of Toronto

participating in support groups, while Green Towns supported community organizational development and leadership training. At the initiation of the project, LHC was focused on awareness creation regarding prevention of spread of HIV/AIDS through training and did not consider livelihood empowerment an urgent matter. Livelihood activities of BM described in the following section, for example, were undertaken quite independently of LHC. The organization was more involved in emergency food distribution than livelihoods and withdrew from the project in December 2007. However, towards the end of the project LHC accepted that they had underrated the need for empowering beneficiaries to support themselves instead of relying on handouts, and offered to work with the project beneficiaries.

In keeping with SEHTUA's commitment to knowledge transfer and capacity building, Egerton University,

the University of Nairobi and the Dairy Goat Breeders Association of Kenya, Nakuru Chapter, joined the team. They provided expertise in animal and human nutrition, gender and group dynamics, livestock–crop interactions, animal health and animal health–environment interactions. The Kenyan and Canadian universities also facilitated access to graduate and undergraduate students to work on distinct aspects of SEHTUA.

Numerous government partners were also important. The Ministries of Agriculture, Livestock and Fisheries Development, and Health and Social Services were crucial in supporting the agricultural and health components respectively. Initial linkages with the MCN on environment-relevant UA policy issues were extended to include the Department of Public Health, a leader in HIV/AIDS prevention and monitoring in conjunction with the National AIDS Control Council (NACC).

Community-based partners

Love and Hope Centre identified the Badili Mawazo Self Help Group (BM) as a group with which to work. Originally affiliated with LHC, BM is an HIV/AIDS psychosocial and welfare development group for PLWHA; the group officially registered as an independent Community Based Organization with the Ministry of Social Services in March 2006. Shortly thereafter, BM partnered with the Presbyterian Church of East Africa (PCEA), Nakuru West Parish, which provided meeting space and other supports. In the words of their founding chairperson:

> The formation of Badili Mawazo was necessitated by the need of PLWHA to come together to help fight stigma and discrimination and form a welfare group through which they can collectively undertake income generating activities to help raise the living standards of its members and their families, who for half of the members also include orphans. This is important as some members lost their previous jobs due to HIV/AIDS, while the majority continue to make their living in the informal sector through precarious and unreliable small businesses (Badili Mawazo, 2008).

To this end they actively participate in skills training and seek partnerships to help households develop diverse and robust livelihood strategies. Since its inception, multiple livelihoods initiatives (besides crops and goat production described subsequently) have been pursued independently of the SEHTUA project, as summarized in Table 2.

The diversity of BM's initiatives combined with the training and support from SEHTUA, represent the

Table 2 | Overview of non-SEHTUA Badili Mawazo livelihood activities

Livelihood activities	Partnerships for implementation
Bead jewellery making and bag weaving	
Bakery project (cakes and mandazi)	
Chicken farming	UNGA Farms, group savings and donations from faith-based organizations in Scotland and Canada
Micro-finance	Kenya Rural Enterprise Program (K-REP)
Soya producing, processing and other food processing at cottage industry level	
Wool spinning	Kenya AIDS NGO Consortium (KANCO)

acquisition of important assets that have contributed to the resilience of the group and its members. BM has been able to survive shocks including: the violence following the December 2007 elections which saw some of their members displaced and others struggle with access to health care and supplies of food and medicines; repeated theft of assets; and crop and livestock failure. Through contacts with local groups (such as ROCK-Bridge Ministries Kenya), partner agencies and visiting scientists and students working with the project, the group purchased a parcel of land to build the Badili Mawazo Greenbank Centre for Hope and Innovation (BMGCHI). The Centre currently includes a meeting house, caretaker house, pit latrine, chicken house and fencing, in addition to room for vegetable gardening and other micro livestock raising initiatives. Several funding proposals have been submitted by the group to both local (Constituency AIDS Control Council – Global fund 2009 and National AIDS Control Council – Total War on AIDS 2009) and international (Stephen Lewis Foundation) organizations.

BM's involvement in chicken farming is indicative of its ability to mobilize multiple supports, and to adapt and apply lessons learned. The original chicken project was funded by a donation from UNGA Farms (a local feed company) to purchase exotic layers as a source of food and potential income through the sale of eggs. After the donation of feed ran out, BM determined that it would not be economically viable to

continue and decided to sell the chickens, bank the money and later start afresh with indigenous chickens. As of the writing of this manuscript, 43 households currently benefit from indigenous chicken production and expansion to additional households is planned.

Organization of SEHTUA

Urban Harvest and Ryerson University as co-principle investigators, plus senior scientists from ILRI and University of Toronto, held meetings at the beginning of the project in Kenya and Canada to agree on respective roles. Initially it was felt that the local co-ordinator for the project should be linked to MCN, to ensure integration of the project in local government. However, the complexity of research for development necessitated a more research-oriented person, so a co-ordinator from Egerton University was hired and a co-ordinating office established in Nakuru in September 2006. SEHTUA management adhered to many of the principles and guidelines for interactive approaches in agriculture innovation in the context of HIV/AIDS proposed by Swaans and colleagues (2006, 2009). In line with the 'farmer first' approach of UH and CGIAR (Scoones and Thompson, 2009), this included a commitment to coalition and capacity building, reflecting respect for local knowledge and different disciplinary backgrounds. Personal commitment on the part of SEHTUA personnel reflected their attachment to BM members and a shared vision of UA for sustainable livelihoods and health.

Engagement of BM executive committee members in decision making around SEHTUA activities strengthened both BM and SEHTUA implementation. On the other hand, when mistrust among BM members was detected, SEHTUA called upon Green Towns to work with UH on Community Organizational Development and Institutional Strengthening (CODIS) training. The training enhanced BM project management, leadership, gender sensitivity and conflict resolution skills, leading to greater stability and organizational growth and increased the competitive ability of some members who were able to take up formal employment.

A commitment to an iterative SEHTUA implementation process allowed flexibility in management team participation and accommodated change in both personal circumstances (maternity leave of a co-principle investigator) and organizational priorities. The project was completed on target despite some significant changes in the policy and funding environment that occurred during the last year of the project. During 2008, as part of a reorganization of the CGIAR, its new visioning document paid very little attention to the impact of urban growth and migration on the levels and location of poverty and rural agriculture and none to agriculture as a food security strategy of the urban poor (CGIAR, 2008). As a result of a reformulation and narrowing of its own research strategy, CIP removed research for development on urban and peri-urban agricultural systems from its research agenda and indicated that it would cease to convene Urban Harvest from 2010 (CIP, 2009). This left the SEHTUA project outside the research priorities of both entities, creating a new funding challenge for going to scale with this project.

SEHTUA implementation

Baseline survey

To better understand the current situation and household practices, a baseline survey generated information on agricultural practices, livelihoods and food and nutrition security of HIV/AIDS-affected households. Agricultural practice questions were based on earlier UH work in Kenya and internationally. Livelihood security adopted the Sustainable Livelihoods Approach, amplified to include outcome measures of age-specific mortality and child illness (de Haan et al., 2002; Andersen et al., 2008). Food security status was assessed using the FANTA Household Food Insecurity Access Scale (Coates et al., 2006) and the household diet diversity scale. Nutrition security was assessed through the dietary intake (via 24-hour recall) and anthropometric measures (weight, height, mid-upper arm circumference, triceps skin-fold measure) of an index child between the ages of 2 and 5 years in the household (Mbugua et al., 2008b).

Participants were drawn from the three main HIV/AIDS support organizations in Nakuru, namely: Catholic Diocese of Nakuru (LHC), ICROSS (International Community for the Relief of Suffering and Starvation), and Family Health International (FHI) Nuru ya Jamii group. The study covered 11 out of the 15 administrative wards in the municipality (Kaptembwo, Shabab, Rhonda, Shauri Yako, Langa Langa, Lake View, Bondeni, Kivumbini, Menengai, and Nakuru East). Exclusion criteria included a household with a child who was very sickly based on current or chronic illness, as this could confound the nutrition security indicators of the household. Recruitment issues were addressed jointly by Love and Hope Centre, Badili Mawazo Executive Committee, the MCN's Public Health Department's HIV/AIDS section which houses the Constituency AIDS Control Committee (CACC) and assisted in linking with ICROSS, and Family Health International based self-help groups.

Results of the baseline survey have been reported elsewhere (Andersen et al., 2008; Cole et al., 2008a; Mbugua et al., 2008b). Briefly, participating households commonly experienced severe food shortage and food insecurity (77 per cent), eviction (37 per cent), hospitalization (34 per cent), job loss (26 per cent), and/or death of an adult (17 per cent). Female-headed households (45.2 per cent of sample) reported more crises (mean 2.83; 95 per cent; CI 2.52–3.13) compared to male headed households (mean 2.10; 95 per cent; CI 1.80–2.40), more illness over the last month (67.1 vs. 57.1 per cent), greater perceived lack of medical care (50 vs. 40 per cent) and less access to land for farming (22.9 vs. 44.7 per cent). The gender differences observed reinforced the focus on inclusion of women in project activities and prompted a gender analysis described in detail elsewhere (Njenga et al., 2009b).

Mean household dietary diversity score in terms of food groups was 8.1 out of a maximum of 15, and was negatively correlated with food insecurity. In terms of frequency of consumption, plant-based foods were generally consumed more frequently than animal source foods (a better source of highly bio-available micronutrients), with the exception of dairy products. Non-vitamin A-rich vegetables were consumed more frequently than Vitamin A-rich and other fruits, although oils and fats (which are required for plant-based sources of vitamin A to be absorbed efficiently) were consumed by almost all participants. Prevalence of stunting (HAZ $< -2SD$) and underweight (WAZ $< -2SD$) was 33.1 and 26.0 per cent, respectively, higher than the national average in the most recent national demographic survey (30.6 and 19.1 per cent respectively; CBS et al., 2004).

Agricultural interventions and nutrition education

Given limited access to land, the project engaged partners (PCEA and the Ministry of Agriculture) to access adequate land. In addition, the project rented an urban parcel of land. Details of each agricultural intervention are described in more detail elsewhere (Njenga et al., 2009b) and are described briefly below. The National AIDS Control Council advocates a three-pronged approach to optimize nutritional outcomes among PLWHA, including medical nutritional therapy, assurance of food and nutrition security, and nutrition education. Final year nutrition undergraduate students from Egerton University conducted the first nutrition training; content was based on the five themes proposed in the Kenyan National Guidelines on Nutrition nd HIV/AIDS. These included the importance of good nutrition, living positively, infection control and food safety, fighting illness through diet, and infant and maternal nutrition in HIV/AIDS (NASCOP, 2006). Subsequent education sessions were provided by the nutritionists from the Provincial General Hospital – Comprehensive Care Centres.

Urban agriculture: horticulture

Eighty households participating in the intervention were divided into two clusters, namely those with and those without their own farming space. Prior to the introduction of the vegetables, participants were trained on vegetable production, utilization and marketing by the agricultural officer in charge of the municipality. Consultations were held between the District Irrigation Officer and the SEHTUA team regarding crop husbandry and irrigating approaches for the vegetable plots. Although one aim was to re-introduce African indigenous vegetables, in keeping with the participatory methodology, participants also chose to grow exotic vegetables. Vegetables grown therefore included: black nightshade, cowpeas, spider plant, amaranthus, and bush okra as well as kales/collards, spinach, cucumber, carrots, onions and beetroots. Inputs for vegetable growing included certified seed from the World Vegetable Centre, fertilizer, manure and implements, together with labour for initial land preparation. Water supply was a challenge, particularly for one larger farm, where irrigation was not available for many crop cycles, reducing yields.

Monitoring and evaluation was implemented to assess participation, use of vegetables and profitability. A qualitative assessment of participants' and former participants' experiences of the intervention was led by a Canadian post doc paired with a Kenyan graduate student. Fifty-two individual semi-structured interviews included current participants (n = 26) and former participants who could be located and were willing to participate (n = 26). Examples of some typical experiences voiced by participants are provided in Table 3. Participation in farm labour was often difficult due to illness among PLWHA and the considerable distance of some farms from participants' living quarters. Poverty among BM members posed a challenge, as agricultural work not directly related to a harvest had a high opportunity cost. For example, people would have to forgo other livelihood activities such as informal selling, in order to go to the farm. In terms of vegetable use, household consumption by participants was important, as was sharing with family and other BM members and sale to neighbours and others, as a source of income. For profitability, a gross margin analysis

Table 3 | **Participants' stories**

Jane*

- A 40-year-old mother of six, once a second-hand clothes dealer.
- Diagnosed as HIV-positive 2 years ago, she spent all her capital on treatment.
- She later joined Badili Mawazo (BM), ... six other women at Manyani, where she learned how to grow vegetables.
- 'Besides taking antiretroviral drugs, the traditional vegetables make me stronger every day.'
- I do not buy vegetables since I started growing my own.
- 'I sell the surplus vegetables and the money I earn lets me meet my children's needs and buy recommended food like eggs, white meat and wheat.' Average sales from BM US$15 and for home consumption worth US$10 per month.

Jackson*

- A father of two, he worked as a guard and a small-time hawker in Nakuru town, but he was getting weaker and weaker.
- 'I had to stay out in the cold all night sometimes on an empty stomach,' he said. 'A medic advised me to quit this strenuous job.'
- As a founder of Badili Mawazo, he has learned to care for the dairy goats.
- 'It changed my social and economic life tremendously.'
- He is happy with what he does and enjoys milking his dairy goat.

*Real names have been concealed for ethical reasons.
Source: Mbugua *et al.*, 2008a.

conducted by a University of Nairobi agriculture student as a field attachment showed that both indigenous and exotic varieties were profitable to grow (Wanjiku, 2007).

Micro livestock: dairy goat keeping

For the micro-livestock intervention, 40 households were selected in a participatory manner by BM. Sensitization workshop participants (see Table 1) came to a consensus to choose dairy goats, in recognition of the need for high-value, low-input livestock that would provide quick returns and respecting concerns regarding potential avian flu. Considerable planning was undertaken by a post-masters' student interning at ILRI (Ferguson, 2007). After an initial analysis of existing goat projects and breeds in Nakuru, Kenyan-Toggenburg were selected. An a priori human health risk scoping assessment was conducted by a

University of Toronto postgraduate student (Chris, 2007) to supplement local consultations with experts to determine the potential sources and ways to mitigate any health risks associated with goat rearing by persons with HIV/AIDS (Kang'ethe *et al.*, forthcoming).

Establishment of fodder banks was the most important activity to be undertaken prior to arrival of the goats so as to ensure availability of sufficient quality feed. After much debate, planting materials comprising sweet potato vines (*Ipomea batata*) and napier grass (*Pennisetum cladistenum)* were selected. They were supplied by the Kenya Agricultural Research Institute, through the National Beef Research Station, Lanet, and five acres of napier and two acres of sweet potato vines were sown.

Goats were procured through the Meru Goat Breeding Association. Prior health screening involved physical examination, collection of blood and faeces for laboratory analysis of *Brucellosis* and *Cryptosporidiosis*. To prepare, goat houses with provisions for feeding area, water, exercise and sleeping were constructed. The Department of Public Health, Pharmacology and Toxicology, University of Nairobi and the Catholic Dioceses of Nakuru provided initial guidance on goat care at a special workshop. Following a one-month acclimatization period, goats were distributed to three clusters, as decided by BM and the SEHTUA team.

Morbidity and mortality among the goats and their offspring was an ongoing challenge. This could partly be attributed to pre-existing conditions (reproductive tract anomaly in one, prior pasteurellosis suspected in several) as well as adverse weather conditions (drought) that reduced fodder yields. Disruption in the scheduled goat care following the post-election violence in early 2008 was a major challenge, and the lack of high quality fodder coupled with the dry season resulted in loss of one goat and eight abortions/stillbirths. Inconsistent participation in goat raising in one cluster due to distance of the farm remained a challenge, as did unequal contributions attributed to sickness and other factors. In response, further training was provided to BM members and the services of a Nakuru-based veterinarian were sought.

In the participatory monitoring and evaluation system, BM members kept daily records and held weekly meetings with the local SEHTUA team. In addition, farm visits by the SEHTUA co-ordinator and occasional visits by the overseas partners enabled the team to address many challenges in a timely fashion. For example, challenges in the regular transportation of fodder or market organic waste for goat feed was resolved through the provision of bicycles to two BM members and paying them a small stipend to regularly provide feed to each goat-keeping cluster.

Professional and researcher capacity building

In addition to the capacity building of BM members described above, professional development of young scientists has been a focus of SEHTUA. Both Kenyan and expatriate students have made important contributions through a combination of field attachments, course work assignments, major research papers, internships, and masters theses (see Table 4). Two post-doctoral fellows contributed expertise and additional funding to the project through the qualitative assessment described under 'urban agriculture' above and in an assessment of infant feeding and HIV/AIDS. The latter project, involving a Kenyan masters student, will examine SEHTUA impact on infant feeding practices in comparison with non-project participants.

Dissemination activities and preliminary results

Badili Mawazo shared their experiences with SEHTUA at an urban agriculture meeting hosted by the Nairobi and Environs Food Security, Agriculture and Livestock Forum (NEFSALF), during World AIDS Day celebrations, and at a special BM Day at the PCEA. A second feedback workshop involving key stakeholders and BM members was organized recently to share successes, challenges and future opportunities. Presentations included participant experiences, partnership development, social and cultural implications on uptake of interventions, infant feeding and HIV/AIDS policies. During discussions, BM participants mentioned the building of social networks, gaining improved self-esteem, increasing money in their household budget, obtaining a regular vegetable supply and accessing goats through their own sweat as benefits of participation.

The workshop's final session included a discussion of the sustainability of the agricultural and livelihood initiatives of BM. Commitment to continued technical support to BM were made by several partners including Egerton University, University of Nairobi, Ministry of Agriculture, LHC, PCEA and ROCK Bridge Ministries. A commitment to pursue the formalization of access to collective farm plots owned by the PCEA and Ministry of Agriculture and for Egerton University and University of Nairobi to support BM in responding to a call for proposals from the National AIDS

Table 4 | **Professional and research capacity building: university students**

Graduate level: thesis	
Egerton University: Human Nutrition	**MSc thesis:** Food and nutrition insecurity status of HIV/AIDS-affected households in Nakuru Municipality
Egerton University: Human Nutrition	**MSc thesis:** Infant feeding, knowledge, attitudes and practices among counsellors and mothers of known HIV status in Nakuru municipality
Egerton University: Sociology	**MA thesis:** Socio-cultural implications on uptake of urban agricultural interventions by HIV/AIDS-affected households: a case of poor urban households in Nakuru Municipality, Kenya
Graduate level: major research paper	
University of Toronto: Anthropology	**MA research paper:** Livelihoods and health status of HIV/AIDS-affected households in Nakuru Kenya
Graduate level: coursework	
University of Toronto: Public Health Sciences	Community Medicine: an assessment of the potential human health risks associated with Urban Harvest Nakuru Project
Graduate level: field attachment	
Egerton University: Institute of Women, Gender and Development Studies	Gender and Development Postgraduate Diploma
Graduate level: internship	
Cornell University: International Agriculture and Rural Development	MPS volunteer internship: stakeholder involvement
Reading University: International and Rural Development	Post-MSc internship: goat intervention
Undergraduate level: field attachment	
Makerere University: Social Work	
University of Nairobi: Agriculture and Veterinary Science	
Undergraduate level: extension course	
Egerton University: Human Nutrition (four students)	
Undergraduate level: internship	
Egerton University: Human Nutrition	
University of Alberta: Human Geography	
University of Nairobi: Veterinary Medicine	
University of Toronto: Environment and Health	

Co-ordinating Committee and Catholic Relief Services were also among the outcomes of the discussion.

Based on preliminary analysis of a repeat survey, increased access to land for agriculture, livestock, technical support services, banking facilities, health facilities and social clubs all seemed to have occurred among participating households. Among those not participating in the interventions, farming activities and participation in social groups had both increased, suggesting some diffusion of knowledge from participants to non-participants. Indicators of household food security improved among participants, while a slight decline among non-participants was noted. Overall, results indicated a positive contribution of SEHTUA to food security and several livelihood capitals among HIV/AIDS-affected households taking part in the agricultural interventions (Njenga *et al.*, 2009a).

Reflection and discussion

In recent years, scholars have produced impressive, cross-disciplinary efforts to evaluate the social, institutional, and ecological outcomes and impacts of agricultural sustainability initiatives (Tiwari *et al.*, 2008). Over a longer period it has been argued convincingly that the sustainability of agri-food systems requires a commitment to build on and integrate the knowledge of food producers and consumers (Prain, 2006; Bawden, 2007; Pralle, 2008; Scoones and Thompson, 2009). Preliminary results from SEHTUA indicate the value of multi-stakeholder investments that bring together affected households, partners from municipal and provincial government and the community, and universities and research institutes. Supporting 'civic' dimensions in action research is not only consonant with agricultural sustainability, but also in keeping with integrated approaches involving diverse sectors in programmes promoting food security and livelihood sustainability with PLWHA (Lyson, 2004; Gillespie and Kadiyala, 2005; Swaans *et al.*, 2006; Panagides *et al.*, 2007).

Projects aiming to link agricultural sustainability and livelihoods are intensive with respect to resources, personnel and financial commitments, both from participants and project partners. Nonetheless, our experience suggests that the investment in a collaborative process can produce desired returns with respect to improved food security, agricultural sustainability and livelihoods, ultimately decreasing the vulnerability of households. BM's improved access to food and income, and increased knowledge through training and capacity building, represent positive changes in forms of capital (natural, human and social) essential to human livelihoods (Rakodi and Lloyd-Jones, 2002). Further, the structured commitments by partners to BM are in keeping with key indicators of sustainability of health promotion interventions (Pluye *et al.*, 2004), which bode well for agricultural livelihoods continuing to be an important resource for PLWHA.

Acknowledgements

First and foremost we want to thank the participants, primarily the Badili Mawazo Self Help Group and particularly the executive without whom there would have been no project, along with the ICROSS and Nuru Ya Jamii Self Help Group members. You were our inspiration. Special thanks to the original proposal partners, in particular T. Randolph (ILRI) and the Love and Hope Centre staff and the many partners listed in Figure 1. Several friends of the project provided invaluable support, including E. Kang'ethe, P. Tuitoek, D. Lee-Smith and P. Munyao. Thanks also to the many students who provided invaluable input and enthusiasm to the project including R. Lavergne, N. Andersen, J. Ferguson, K. LaFleche, L. Kramer, K. Chrichton-Struthers, A. Cherobon, G. Muirithi, H. K. Njunga, M. R. Kiome, J. K. Chege, E. M. Mwanja, A. Chris, W. Gachie, E. Wamuhu, J. M. Kariuki, M. G. Wanjiku, I. Kigen, C. Mwai and G. Muiga. Thanks finally to the Canadian International Development Agency for the primary funding, and additional funding from the Canadian Institutes for Health Research and Canadian Research Chair program.

Note

1. The International Food Policy Research Institute, one of the CGIAR Centres based in Washington, has also conducted research on food systems and urban poverty.

References

AED (Academy for Educational Development), 2008, *Nutrition, Food Security and HIV – A Compendium of Promising Practices* [available at www.fantaproject.org].

Andersen, N., Mbugua, S., Sellen, D., Cole, D., Karanja, N., Yeudall, F., Prain, G., Njenga, M., 2008, 'Applications of the sustainable livelihoods framework to assess households affected by HIV/AIDS in Nakuru, Kenya', in: *XVII International AIDS Conference Abstract Book*, Vol. 1 (p. 253) [available at www.aids2008.org].

Badili, Mawazo, 2008, 'Stephen Lewis Foundation Grant Proposal', Unpublished project document available on request from the corresponding author.

Bawden, R. J., 2007, 'A paradigm of persistence: a vital challenge for the agricultural academy', *International Journal of Agricultural Sustainability* 5 (1), 17–24.

Bebbington, A., 1999, 'Capitals and capabilities: a framework for analyzing peasant viability, rural livelihoods and poverty', *World Development* 27 (12), 2021–2044.

Boischio, A., Clegg, A., Mwagore, D. (eds), 2006, *Health Risks and Benefits of Urban and Peri-urban Agriculture and Livestock (UA) in Sub-Saharan Africa*, Urban Poverty and Environment Series Report #1, International Development Research Centre, Ottawa.

Castleman, T., Seumo-Fosso, E., Cogill, B., 2003, *Food and Nutrition Implications of Antiretroviral Therapy in Resource Limited Settings*, Academy for Educational Development (AED), Food and Nutrition Technical Assistance Project (FANTA), Washington, DC.

CBS (Central Bureau of Statistics), Ministry of Health and ORC Macro, 2004, *Kenya Demographic and Health Survey 2003*, CBS, MoH and ORC Macro, Calverton, MD.

CGIAR, 2008, *Visioning the Future of the CGIAR*, Report of Working Group 1 (Visioning) to the Change Steering Team of the CGIAR, 6 June, 2008, CGIAR, Washington, DC [available at www.cgiar.org/changemanagement/pdf/wg1_VisioningChange_Final_Report_June6.pdf].

Chris, A., 2007, 'An Assessment of the Potential Human Health Risks Associated with Urban Harvest Nakuru Project', Unpublished project document available on request from the corresponding author.

CIP, 2009, *The Strategic and Corporate Plan of the International Potato Center, 2009–2018*, CIP, Lima.

Coates, J., Swindale, A., Bilinsky, P., 2006, *Household Food Insecurity Access Scale (HFIAS) for Measurement of Household Food Access: Indicator Guide*, Food and Nutrition Technical Assistance (FANTA) Project, AED, Washington, DC.

Cohen, B., 2004, 'Urban growth in developing countries: a review of current trends and a caution regarding existing forecasts', *World Development* 32 (1), 23–51.

Cole, D., Andersen, N., Samuel, N., Karanja, N., Yeudall, F., Njenga, M., Sellen, D., Prain, G., SEHTUA, 2008a, 'SEHTUA: community based action research project to strengthen livelihood and food security for households affected by HIV/AIDS through urban agriculture', in: *XVII International AIDS Conference Abstract Book*, Vol. 1 (p. 319) [available at www.aids2008.org].

Cole, D., Lee-Smith, D., Nasinyama, G. (eds), 2008b, *Healthy City Harvests: Generating Evidence to Guide Policy on Urban Agriculture*, CIP/Urban Harvest and Makerere University Press, Lima, Peru.

De Haan, A., Drinkwater, M., Rakodi, C., Westley, K., 2002, *Methods for Understanding Urban Poverty and Livelihoods* [available at www.livelihoods.org].

Dennery, P., 1996, 'Urban food producers' decision-making: a case study of Kibera, City of Nairobi, Kenya', *African Urban Quarterly* 11 (2 & 3), 189–200.

Ferguson, J., 2007, 'Urban Harvest Nakuru Project. Dairy goat Report', Unpublished project document available on request from the corresponding author.

Foeken, D., 2006, *'To Subsidise My Income' – Urban Farming in an East-African Town*, Brill, Boston, MA & Leiden.

Foeken, D., Owuor, S., 2000, *Urban Farmers in Nakuru, Kenya*, ASC Working Paper No. 45, Africa Studies Centre, Leiden, and Centre for Urban Research, University of Nairobi.

Gillespie, S., Kadiyala, S., 2005, *HIV/AIDS and Food and Nutrition Security. From Evidence to Action. Food Policy Review 7*, International Food Policy Institute, Washington, DC.

Hooton, N., Lee-Smith, D., Nasinyama, G., Romney, D., Atukunda, G., Azuba, M., Kaweesa, M., Lubowa, A., Muwanga, J., Njenga, M., Young, J., 2007, *Championing Urban Farmers in Kampala. Influences on Local Policy Change in Uganda. Process and Partnership for Pro-poor Policy Change*, ILRI Research Report No. 2, ILRI, Nairobi (Kenya).

Kiarie, S., 2009, 'HIV and AIDS Policy in Kenya – Nakuru Municipal Council', Unpublished project document available on request from the corresponding author.

Kang'ethe, E. K., Njehu, A., Karanja, N., Njenga, M., Gathuru, K., Karanja, A., forthcoming, 'Benefits and selected health risks of urban dairy production in Nakuru, Kenya', in: G. Prain, N. KaranjaD. Lee-Smith (eds), *Africa Urban Harvest: Agriculture In and Around African Cities 2002–2006*, IDRC/Urban Harvest/CIP.

Karanja, N., Njenga, M., Gathuru, K., Karanja, A., Munyao, P., forthcoming, 'Crop–livestock–waste interactions in Nakuru, Kenya', in: G. Prain, N. Karanja, D. Lee-Smith (eds), *Africa Urban Harvest: Agriculture In and Around African Cities 2002–2006*, IDRC/Urban Harvest/CIP.

Lee-Smith, D., Mutsembi, M., Lamba, D., Gathuru, P.K., 1987, *Urban Food Production and the Cooking Fuel Situation in Urban Kenya*, National Report, Mazingira Institute, Nairobi.

Loevinsohn, M., Gillespie, S., 2003, *HIV/AIDS, Food Security and Rural Livelihoods: Understanding and Responding*, RENEWAL Working Paper No. 2, May [available at www.ifpri.org/renewal/pdf/RENEWALWP2.pdf].

Lyson, T.A., 2004, *Civic Agriculture: Reconnnecting Farm, Food and Community*, Tufts University Press, Medford, MA.

Masariala, W., 2007, 'The poverty–HIV/AIDS nexus in Africa: a livelihood approach', *Social Science and Medicine* 64, 1032–1041.

Maxwell, D., 1995, 'Alternative food security strategy: a household analysis of urban agriculture in Kampala', *World Development* 23 (10), 1669–1681.

Maxwell, D., Ahiadeke, C., Levin, C., Armar-Klemesu, M., Zakariah, S., Lamptey, G. M., 1999, 'Alternative food security indicators: revisiting the frequency and severity of 'coping strategies', *Food Policy* 24 (2), 411–429.

Mbugua, S., Karanja, N., Njenga, M., 2008a, 'Gardens of hope for victims of HIV/AIDS in Nakuru, Kenya', in: *Urban Grown* (Newsletter of the Kansas City Centre for Urban Agriculture) [available at www.kccua.org/urbangrown/ug-10-08.htm].

Mbugua, S., Andersen, N., Tuitoek, P., Yeudall, F., Sellen, D., Karanja, N., Cole, D., Njenga, M., Prain, G., SEHTUA, 2008b, 'Assessment of food security and nutrition status among households affected by HIV/AIDS in Nakuru Municipality, Kenya', in: *XVII International AIDS Conference Abstract Book*, Vol. 1 (p. 493) [available at www.aids2008.org].

McCarney, P., Stren, R. E. (eds), 2003, *Governance on the Ground: Innovations and Discontinuities in Cities of the Developing World*, Johns Hopkins University Press, Baltimore, MD.

Mukisira, E., 2005, 'Opening Speech', in: G. Ayaga, G. Kibata, D. Lee-Smith, M. Njenga, R. Rege (eds), *Prospects for Urban and Peri-urban Agriculture in Kenya*, Urban Harvest–International Potato Centre, Lima.

NASCOP (National AIDS/STD Control Programme), 2003, *Kenya National HIV/AIDS Strategic Plan 2005/6 to 2009/10. A Call to Action* [available at www.aidskenya.org].

NASCOP (National AIDS/STD Control Programme), 2006, *Kenyan National Guidelines on Nutrition and HIV/AIDS* [available at www.aidskenya.org].

NASCOP (National AIDS/STD Control Programme), 2008, *Kenya AIDS Indicator Survey 2007 KAIS Preliminary Report* [available at www.aidskenya.org].

Njenga, M., Karanja, N., Gathuru, K., Mbugua, S., Fedha, N., Ngoda, B., 2009a, 'The role of women-led micro-farming activities in combating HIV/AIDS in Nakuru, Kenya', in: A. Hovorka, H. de Zeeuw, M. Njenga (eds) *Women Feeding Cities: Mainstreaming Gender in Urban Agriculture and Food Security*, Practical Action Publishing, Rugby, UK.

Njenga, M., Karanja, N., Mbugua, S., Muriithi, G., Cherobon, A., SEHTUA, 2009b, *Combating HIV/Aids in Urban Communities through Food and Nutrition Security: The Role of Women Led Micro-Livestock Enterprises and Horticultural Production in Nakuru Town*, Feedback from Workshop held on 11 June, Jumuia Guest House-Nakuru, Kenya.

Njenga, M., Romney, D., Karanja, N., Gathuru, K., Kimani, S., Carsan, S., Frost, W., forthcoming, 'Recycling nutrients from organic wastes in Kenya's capital city', in: G. Prain, N. Karanja, D. Lee-Smith (eds), forthcoming, *African Urban Harvest: Agriculture In and Around African Cities, 2002–2006*, IDRC/Urban Harvest/CIP.

Panagides, D., Graciano, R., Atekyereza, P., Gerberg, L., Chopra, M., 2007, *A Review of the Nutrition and Food Security in HIV and AIDS Programs in Eastern and Southern Africa*, Equinet discussion paper number 48 [available at www.equinetafrica.org].

Pluye, P., Potvin, L., Denis, J.L., 2004, 'Making health programs last: conceptualizing sustainability', *Evaluation and Program Planning* 27, 121–133.

Prain, G., 2006, 'Participatory technology development for urban agriculture: collaboration and adaptation along the urban–rural transect', in: R. van Veenhuizen (ed.), *Cities Farming for the Future – Urban Agriculture for Green and Productive Cities*, RUAF Foundation, IDRC and IIRR, Leusden.

Prain, G., Karanja, N., Lee-Smith, D. (eds), forthcoming, *African Urban Harvest: Agriculture In and Around African Cities, 2002–2006*, IDRC/Urban Harvest/CIP.

Pralle, J., 2008, 'Agricultural sustainability: concepts, principles and evidence', *Philosophical Transactions of the Royal Society B* 363, 447–465.

Rakodi, C., Lloyd-Jones, T., 2002, *Urban Livelihoods: A People-Centred Approach to Poverty Reduction*, Earthscan/DFID, London.

Scoones, I., Thompson, J. (eds) 2009, *Farmer First Revisited. Innovation for Agricultural Research and Development*, Practical Action Publishing, Rugby, UK.

SEHTUA (Sustainable Environments and Health Through Urban Agriculture), 2006, *Combating HIV/AIDS in Urban Communities Through Food and Nutrition Security*, Report of the Sensitization Workshop organized by Urban Harvest, CIP, SSA, Ryerson University and University of Nairobi.

Swaans, K., Broerse, J., Bunders, J., 2006, 'Agriculture and HIV/AIDS: a challenge for integrated and interactive approaches', *Journal of Agricultural Education and Extension* 12 (4), 231–247.

Swaans, K., Broerse, J., Meincke, M., Mdhara, M., Bunders, J., 2009, 'Promoting food security and well-being among poor and HIV/AIDS affected households: lessons from an interactive and integrated approach', *Evaluation and Program Planning* 32, 31–42.

Tevera, D. S., 1996, 'Urban agriculture in Africa: a comparative analysis of findings from Zimbabwe, Kenya and Zambia', *African Urban Quarterly* 11 (2 & 3), 181–188.

Tiwari, K. R., Nyborg, I. L. P., Sitaula, B. K., Paudel, G. S., 2008, 'Analysis of the sustainability of upland farming systems in the Middle Mountains region of Nepal', *International Journal of Agricultural Sustainability* 6 (4), 289–306.

Wanjiku, M. G., 2007, *Sustainable Environments and Health Through Urban Agriculture (SEHTUA) Urban Harvest-Nakuru Project*, A field attachment report submitted in partial fulfillment for the requirements of the degree of Bachelor of Science in Agriculture, College of Agriculture and Veterinary Sciences. University of Nairobi.

Putting the culture back into agriculture: civic engagement, community and the celebration of local food

Jennifer Sumner[1]*, Heather Mair[2] and Erin Nelson[3]

[1] Adult Education and Community Development Program, OISE/University of Toronto, 252 Bloor Street West, Toronto, Ontario M5S 1V6
[2] Department of Recreation and Leisure Studies, University of Waterloo, 200 University Avenue West, Waterloo, Ontario N2L 3G1
[3] Rural Studies Program, University of Guelph, Guelph, Ontario N1G 2W1

This paper reports on the case study of a community-supported agriculture (CSA) farm in south-western Ontario, Canada. As an exemplar of urban agriculture, Fourfold Farm CSA operates from an alternative agriculture paradigm and is built upon the socio-ecological practices of civic engagement, community and the celebration of local food. Analysis of in-depth, key informant interviews with members of the CSA as well as the co-founders reveals the extent to which the farm is much more than a source of healthy, organic food. The paper outlines the ways the CSA operators and their members articulate a deeper endeavour to link urban food consumers with food producers through cultural activities. The discussion concludes with a call for more social research in agriculture as well as a broader effort to articulate the ways urban agriculture can contribute to putting the culture back into agriculture and creating sustainable systems of farming.

Keywords: civic engagement, community, community-supported agriculture (CSA), local food, urban agriculture

Introduction

Agriculture can be understood as a linked, dynamic social-ecological system based on the extraction of biological products and services from an ecosystem, innovated and managed by people, encompassing all stages of production, processing, distribution, marketing, retail, consumption and waste disposal (McIntyre *et al.*, 2009). Begun 12,000 years ago in the Fertile Crescent of the Middle East, agriculture has evolved into an industrial giant in the 21st century. However, along the path of its recent evolution, something crucial has been lost. In the words of Jules Pretty (2002, p. xii):

In the earliest surviving texts on European farming, agriculture was interpreted as two connected things: *agri* and *cultura*, and food was seen as a vital part of the cultures and communities that produced it. Today, however, our experience with industrial farming dominates, with food now seen simply as a commodity, and farming often organized along factory lines. The questions I would like to ask are these. Can we put the culture back into agri-culture without compromising the need to produce enough food? Can we create sustainable systems of farming that are efficient and fair and founded on a detailed understanding of the benefits of agro-ecology and people's capacity to cooperate?

This paper aims to develop answers to Pretty's probing questions. In addition, it seeks to add to a growing effort, well articulated by McDonald (2005, p. 71), to broaden understandings of the socio-cultural relationships underpinning agriculture by 'exploring the complex ways that people conceptualize, give meaning to, and organize around agriculture broadly conceived'. The paper begins by developing the theoretical framework for the study, building on Beus and Dunlap's (1990) description of two paradigms shaping modern agriculture: conventional and alternative. This is followed by a short discussion of urban agriculture and community-supported agriculture

*Corresponding author. Email: jennifer.sumner@utoronto.ca
INTERNATIONAL JOURNAL OF AGRICULTURAL SUSTAINABILITY 8 (1&2) 2010
PAGES 54–61, doi:10.3763/ijas.2009.0454 © 2010 Earthscan. ISSN: 1473-5903 (print), 1747-762X (online). www.earthscan.co.uk/journals/ijas

earthscan

(CSA). Next, a case study of urban agriculture in southern Ontario is presented to set the scene for considerations of civic engagement, community and the celebration of local food – all expressions of cultural practices associated with the alternative paradigm. The paper concludes by revisiting Pretty's questions.

Theoretical framework: the two paradigms of agriculture

Beus and Dunlap (1990) have put forward an argument for two socio-cultural paradigms influencing agriculture: the conventional paradigm of large-scale, highly industrialized agriculture, and an increasingly vocal alternative agriculture movement, which advocates major shifts toward a more ecologically sustainable agriculture. They sought to clarify and synthesize the core beliefs and values underlying these two approaches to agriculture and outlined six major dimensions: (1) centralization vs. decentralization, (2) dependence vs. independence, (3) competition vs. community, (4) domination of nature vs. harmony with nature, (5) specialization vs. diversity, and (6) exploitation vs. restraint. To this list, Chiappe and Flora (1998) added two more dimensions to the alternative agriculture approach – quality family life and spirituality – while Sumner (2003) added a third – conscious resistance to corporatization. Although there is undoubtedly a range of approaches to agriculture, these two paradigms not only illustrate the socio-cultural tensions within agriculture today, but also help us to understand the vital role of culture and values in agriculture as well as their connection to sustainability. As Jackson has argued, 'Our cultural values will be paramount in determining the outcome of the [paradigmatic] conflict' (Jackson, 1987; cited in Beus and Dunlap, 1990, p. 596).

In this way, culture represents a signpost that differentiates the two paradigms of agriculture – a fork in the road that either suppresses or supports 'one of the main elements of every social system' (Johnson, 2000, p. 73). Initially associated with skilled human activities through which non-human nature was encompassed and transformed – as in agriculture (Cosgrove, 2000) – culture is now understood as 'the distinctive ideas, customs, social behaviour, products or way of life of a particular society, people, or period' (OED Online, 2009). Around the world, these distinctions have been nurtured through policies and practices, not only in the valorization of skilled human activities implicit in, for example, the management-heavy prescriptions of organic or biodynamic agriculture, but also in the communities, ideas, customs, social behaviours, products and ways of life associated with alternative forms of agriculture.

Urban agriculture and community-supported agriculture

As an example of the alternative paradigm, urban (and peri-urban) agriculture has been described as 'agriculture occurring within and surrounding the boundaries of cities throughout the world and includes crop and livestock production, fisheries and forestry, as well as the ecological services they provide' (McIntyre et al., 2009, p. 288). Urban agriculture is undertaken by residents in an effort to take control of food security, social ills and environmental degradation in their communities by providing food, jobs, environmental enhancement, education, beautification, inspiration and hope (Bourque, 2000). Around the world, it is estimated that 800 million city dwellers, including some in industrialized countries, use their agricultural skills to feed themselves and their families (Millstone and Lang, 2003). Pretty (2002) reports that in some Latin American and African cities, up to one-third of vegetable demand is met by urban production; in Hong Kong and Karachi it is about half; in Shanghai more than four-fifths; and in Cuba, it is a central part of the whole country's food security. In large American cities,

> a nascent but ambitious movement is already under way to cultivate a new urban agriculture, with operations ranging from backyard and rooftop gardens to restaurant salad gardens to large community farms located in greenbelts and reclaimed industrial areas that produce orchard fruit, vegetables, honey, even livestock and farmed fish (Roberts, 2008, p. 308).

In Canada, urban agriculture is a small but growing example of the alternative paradigm that is practised in city allotments, community gardens, SPIN (Small Plot Intensive) farming, remedial projects and CSA programmes. As the current study makes clear, CSAs exemplify both urban agriculture and efforts to put the culture back into agriculture.

CSA (also known as community-shared agriculture) is 'an arrangement whereby a group of people, one of whom is a farmer, agree to share the costs and products of a seasonal vegetable garden' (Fieldhouse, 1996, p. 43). Consumers relate to the farmers in one of three main ways. They may be shareholders, members or subscribers and while each involves paying a set amount to the farmer in return for food, consumers may also engage in various activities on and off the farm and contribute to farm operations at various levels (Abbott Cone and Kakaliouras, 1995). Van En (1995) explains that the idea of CSA originated in Japan in the 1960s when homemakers began noticing an increase in imported foods, the consistent loss of farmland to development and the migration of farmers to the cities. She describes how a group of women approached a local farm family with an idea to address these issues and provide their families with fresh fruits and vegetables. A contract was drawn up and 'the "teikei" concept was born, which translated literally means partnership, but philosophically means "food with the farmer's face on it"' (p. 1). CSAs began appearing in North America in 1986 and within a decade it was estimated that there were 'at least 566 CSA projects in the U.S. and Canada, and hundreds more around the world' (Kolodinsky and Pelch, 1997, p. 130). There are currently more than 1300 registered CSA farms across North America (Robyn Van En Centre, 2009) expressing a diversity that is captured in Henderson (2007, pp. 8–9): 'Like grapes or garlic, CSA takes on the flavour, bouquet, and integrity of where it grows, becoming appropriately adapted to each unique situation.'

In their study of a CSA farm in Michigan, DeLind and Ferguson (1999, p. 191) argue that CSA 'provides a social and economic alternative to the conventional, large-scale, corporately managed food system'. They describe how farmers gain a reliable market and the financial (as well as labour and social) support of members prior to each season, while members receive weekly shares of fresh, locally grown produce 22–52 weeks of the year, depending on the region. As a local institution 'designed to share the risks and rewards of farming' (p. 191), CSA can be considered as a 'form of direct marketing of agricultural products that can be an important facet of a more sustainable, locally based food system' (Kolodinsky and Pelch, 1997, p. 129).

In their paper examining the cultural role of two CSA farms in Illinois, McIlvaine-Newsad *et al.* (2004) point out that, 'CSA not only seeks to redefine agriculture, as we know it, but also attempts to establish personal relationships between farmers, consumers and specific places, thus fostering a sense of community...' (p. 149). In addition, CSA farms aim to meet the broad goals of sustainable agriculture by: 'providing a wide range of healthy, locally-grown food; reconnecting consumer with producer; and fostering a sense of local environmental and human stewardship' (McIlvaine-Newsad *et al.*, p. 153). For this reason, CSA farms can provide researchers with a working model for addressing Pretty's questions: how can we put the culture back into agriculture while contributing to the overall production of food and creating sustainable systems of farming?

Methodology and methods: a case study of Fourfold Farm CSA

Case studies, according to Babbie (2001), are useful insofar as they allow researchers to limit their attention to a particular social phenomenon in a particular area. In his discussion of relevant situations for different research strategies, Yin (2003) argues that case studies foster a consideration of research questions that ask 'how' and develop an understanding of contemporary events or phenomena. Stake (2005) makes a distinction between *intrinsic* case studies (those where a researcher undertakes to understand the specifics of a case and not to generalize) and *instrumental* case studies (those where a researcher undertakes to support or build understanding of general phenomena by looking in depth at one typical example). As this study seeks to build our understanding of the cultural role of CSAs, we follow an instrumental case study approach. The questions guiding the study then were informed by the literature outlined above and centred on the goal of understanding how culture is articulated and integrated in the alternative paradigm through urban agriculture, and in particular, through the operation of this CSA.

The methods selected for the study included qualitative interviews and participant observation. As with other studies of urban agriculture, including community gardens (see for example, Chung *et al.*, 2005) and CSA farms (see for example, Abbott Cone and Kakaliouras, 1995; McIlvaine-Newsad *et al.*, 2004) these methods were chosen to elicit rich, detailed data. To generate interest in the study and identify potential participants, the researchers first approached the owners of the CSA and explained the study. The owners offered to advertise the project through their membership email list and interested participants were encouraged to contact the researchers to schedule an interview. As a result, in-depth, semi-structured, one-on-one interviews were conducted with six CSA members. Smith (2005) suggests that a small number of participants in a study can provide an entry point into understanding larger networks of relations. By focusing on the cultural aspects of CSAs, this study provides an entry point into answering Pretty's (2002) questions about putting the culture back into agriculture and re-establishing the relationships among people, land and nature in a more sustainable manner. The six members were asked to reflect on their reasons for being part of the CSA, their sense of the place of the CSA in the broader community and their views on the role of culture (mainly evidenced through social gatherings and community events) in its overall operation. The interviews, lasting from 30 to 90 minutes in length, were audio-recorded and transcribed. The data were analysed using open coding and axial coding techniques. As Charmaz (2006) describes, open or line-by-line coding pulls the data apart into pieces that are initially categorized into descriptive words or phrases. The next step, axial coding, brings the data back together into a coherent, descriptive or explanatory whole. Following Strauss (1987), these techniques allowed the data to be increasingly scrutinized in order to produce appropriate concepts and themes to guide the discussion of the findings and insights into our understanding of the case.

In addition to the interviews with members, one unstructured, in-depth, dyad interview was conducted with the couple who founded and operate the CSA. They were asked to outline the history of the CSA and reflect on the challenges of operating the CSA as well as to discuss future strategies for working with their local community. The data from this interview were used as a form of background information and provide context to the other interviews.

Fourfold Farm CSA

Fourfold Farm CSA was launched in 2000 by Mark Ross and Sandra Moerschfelder. Located within the urban shadow of Guelph, Ontario, Fourfold is a Demeter-certified, biodynamic farm with 37 workable acres, including 10 acres of vegetables, flowers, herbs, cover crops, laying hens, ducks and a small herd of cattle. The farm also has a booth at the year-round farmers' market in Guelph. Like many organic and biodynamic farms, Fourfold mentors young people by hosting CRAFT (Collaborative Regional Alliance for Farmer Training) interns and students from Canada World Youth. It is also the home base for the Society for Biodynamic Farming and Gardening in Ontario, and publishes the Society's seasonal newsletter.

When Fourfold Farm CSA began, it delivered food boxes to both Guelph and the nearby city of Kitchener-Waterloo, and had a farm pick-up as well. At its height, the CSA supplied 120 households during the growing season with biodynamically grown food. Over time, however, the owners found they were spread too thinly and had little time left for themselves or their children. Consequently, they scaled back their operation to 100 households within the Guelph area, with consumers picking up their boxes of produce from a designated place within the city each week. In the future, the

owners plan to make further changes by opening up more space to deal with some necessary cleaning-up tend to loose ends and build infrastructure – all to carry them more sustainably into the future and help them develop a more conscious accounting and appreciation of the quality of community that gathers around the farm.

Over the years, Fourfold has been at the centre of a range of cultural activities. It hosts Treble in the Fields, an annual music festival, and shows a farm-related film in the barn as part of the annual Guelph International Film Festival. It has also organized strawberry socials and pesto-making workshops. In keeping with its biodynamic roots in the teachings of Rudolph Steiner, Fourfold maintains an ongoing relationship with the local Waldorf School, holding its annual Shepherd's Play in the barn in mid-December and organizing 'Wallypalooza' – a fundraiser for the school. Kindergarten students from the Waldorf School come out to the farm once a week to experience farm life and senior students help with the harvest. This relationship is consistent with Henderson's (2007, p. 158) finding that 'several biodynamic CSAs have close relationships with Waldorf schools and regularly host programs for the children'.

Notwithstanding the diversity in CSAs, Fourfold Farm CSA shares some typical characteristics with many CSAs in North America. For example, consistent with Lass et al.'s (2001) national survey, the owners of Fourfold Farm are youthful, well educated, less reliant on non-farm income than conventional farmers, and use a diverse combination of labour. Like other CSA farms, the farm is small and produces organically, and the CSA operation is just one way that these farmers market their products – they also sell their produce at the farmers' market.

In their study of CSA in the Midwest United States, Tegtmeier and Duffy (2005) found, among other things, that the typical farmer was 45 years old and most likely a college graduate. In addition, the average farm was just over 30 acres and the CSA had been operating for more than five years. These findings mirror the case of Fourfold Farm CSA. Unlike Tegtmeier and Duffy, however, Fourfold Farm CSA serves more than 33 members, raises animals as well as produce, and (besides the two owners themselves) does not use family members to provide the majority of labour.

Discussion

Data analysis revealed three key themes highlighting the ways culture is a central part of the CSA under study: civic engagement, community and the celebration of local food.

Fostering civic engagement

Fostering civic engagement is one of the predominant themes emanating from the data analysis. As the following quotations indicate, participants described their deliberate efforts to join the CSA as a reflection of their growing concern about access to healthy food and the desire to support local farmers.

> The idea of having your food come from somewhere local, in season and organic I think are fairly important. Knowing [the co-founders], ... and now contributing to their livelihood and their way of life is sort of an interesting way of looking at it, rather than just contributing to Zehrs [large grocery chain in south western Ontario] ... (participant #6).

> Obviously helping the local farmer. I've got farm history in my family as well so that was pretty important. Yeah, getting produce and vegetables and products that are organic, without pesticides, which aren't hurting the land, the production of which is not hurting the land (participant #1).

> ... supporting people who are growing organically is a really good thing to do. ... We make organic and conventional nut butter, so that whole idea has been part of my consciousness for a long time, so belonging to a CSA is another way of helping that style of agriculture grow (participant #2).

> I just feel I've been concerned about the whole GMO thing for several years, passionately. And I feel that it's a key issue that strikes right at the heart of agriculture. It's the core of humanity. ... I think all those issues need to be addressed. What do we do about GMOs? How have we allowed that to happen? What is the role of the consumer in relation to that? Those are my questions. I want to know why that is. How did it come about that we haven't cared enough to want to force the government – that we've allowed them to build this structure? I find it incredibly upsetting. And CSA is just all movement in the right direction (participant #3).

These comments support the findings of other CSA research. For example, Tegtmeier and Duffy (2005, p. 5) report that, 'CSA may spur local, civic involvement by energizing environmental initiatives, preservation of open and rural landscapes and other community-building activities.' Similarly, Lass *et al.* (2001) found that being involved in a CSA helped farmers improve their community involvement. Abbott Cone and Kakaliouras (1995) denote CSA as a social movement whereby a new institutional arrangement is created 'that addresses multiple societal needs – healthy food, healthy land and healthy social relationships' (p. 29). Their research into four CSA farms in the United States revealed that members articulated their commitment to the farms in moral terms – as part of their moral responsibility to care for the land, communities and food producers. Lyson (2004, 2005) argues that civic agriculture embeds agricultural and food production in the community; it involves 'community-based agriculture and food production activities that create jobs, encourage entrepreneurship and strengthen community identity' (2005, p. 96). While he often refers most directly to networks of producers, he argues CSAs are visible forms of such enterprises, helping to create active food citizens. Further, he points out that no matter the form, these enterprises are examples of 'local problem-solving activities organised around agriculture and food' (p. 98).

Building community

Although a consistent theme, the notion of building community emerged from the interviews in a variety of complex ways. In some cases, participants expressed direct enjoyment in just being part of the community experience of farming.

> I love the community aspect. We'll go up there [to the farm] and the kids get to play with [the children of the founders] and see all the beautiful food that's just been picked and it's more immediate and I want them to know where food comes from ... To have that it's more of a relationship with the people who grow your food and then the possibilities of community centred around supporting farmers (participant #5).

> And when we first joined, we went to a potluck supper there and there's a barn dance, and I know they show movies and stuff through the film festival. And another amazing thing is every year they host the Shepherd's Play and to see that in a barn ... you just can't beat it in a barn, so there are all these people that my kids know and it's at Fourfold where their food is grown and my son's grade 3 teacher is playing Mary and it just really feels like being part of a village and people from other parts of your life are weaving into other things and it really breaks down that compartmentalization that happens in modern culture and makes you really feel like part of this village (participant #4).

Participant #3 expressed it differently:

> I feel part of a community for sure, but I don't necess-
> arily find that to be an active external community. To
> me, it's more of an inner thing. . . . It's not like we're
> all friends or we're a club or a clique . . . I'm actually
> looking for community based on need, on values. To
> use that word 'organic' it's something that's actually
> real . . . So in that sense, I do very much feel part of a
> community, and even if it's not active and external,
> you know that it's implicit that you are connected
> to these people. They know you. It becomes quite
> an important network in a deeper way.

In another vein, some participants indicated they didn't feel especially connected to the social events organized by the CSA. As participant #2 notes:

> . . . I've been a couple of times to the all-day music
> events and it's OK, but it's not my thing. . . . I feel
> that the young ones probably get a lot more of the
> social, community culture out of belonging to a
> CSA. For me, that's not what I'm looking for.
> These guys aren't going to be my best friends. . . .
> for me it's much more about supporting [the foun-
> ders] personally and supporting organic agriculture.

In other instances, the essence of belonging to a community through the CSA farm was articulated through the theme of ownership in a number of interviews. As participant #1 argues:

> I think it was good for them [participant's children] to
> see the vegetable covered in dirt, to help with
> washing and then we'd talk about them and say,
> 'Oh that's nice. It's from our farm.' And we'd feel
> more of a sense of ownership over that.

Although scholars have been studying the idea of community for over 150 years, there is still a great deal of disagreement among them. Such disagreement highlights the complexities and nuances associated with the term, which are reflected in the interviews. There is little doubt, however, that community is linked to culture, as affirmed by Bell and Newby (1961, in Johnston, 2000, p. 101): 'community membership involves "a matter of custom and of shared modes of thought or expression, all of which have no other sanction than tradition".' The complexity of community is not lost on CSA researchers. McIlvaine-Newsad *et al.* (2004) for instance, argued that community has often been understood rather naively or even taken for granted, while DeLind argues that the goal of building community within CSAs may exist more as a metaphor than a fact (2003, as cited in McIlvaine-Newsad *et al.*,

2004). That said, evidence of building community abounds in CSA research. For example, Lass *et al.* (2001, p. iii) found that 'many CSA farms – 73.5 per cent of the farms that responded – organized social and educational events for their shareholders and their communities'. Tegtmeier and Duffy (2005, p. 5) describe CSA as 'a novel marketing and community-building concept' and argue that, at its best, CSA offers members 'a community-building connection with farmers, neighbours and landscapes'. And Henderson (2007) mentions the benefits of CSA to communities and pushes the concept one step further by discussing agriculture-supported communities.

Celebrating local food

A third theme emanating from the data analysis is celebration. For example, participant #4 described:

> . . . the one they have annually that's called Treble in
> the Fields and you go out there and there's food and
> activities for the kids, like fishing for frogs or playing
> in the barn and it's a lovely outdoor experience, and
> it's eclectic and they blur Wallypalooza with Treble
> in the Fields and they have bellydancing and one
> year they had the Guelph seniors' swing band so
> these seniors doing Glen Miller in tuxes were there
> in the fields.

Celebration can also be more personal and spiritual, as participant #3 suggests:

> I would say to my daughter, 'you know, this is from
> the farm', and I would always draw that connection
> and literally feel that connection. And I think just
> the pleasure of that, even if you're not totally
> loving one thing on your plate, it's like this is from
> the farm so we're using it and it's an amazing taste.
> You know you're having an incredibly sacred con-
> nection to what you eat.

In addition to these forms of celebration, Fourfold Farm CSA has held strawberry socials, potluck dinners and sent out newsletters. Overall, in terms of putting the culture back into agriculture, participant #3 summed it up well:

> . . . that's what I love about Fourfold. That there is
> something, it's that overlay of culture. It's wonderful.

Celebration is one of the clearest expressions of culture, as evidenced by the ceremonial or festive life of a community. Research on CSA farms indicates a common celebratory trend. For example, Lass *et al.* (2001) found that almost three-quarters of the CSAs they surveyed organized events, including potluck dinners and arm tours. Other researchers report such cultural events as festivals, potluck suppers, food demonstrations and the creation of a newsletter (see Abbott Cone and Kakaliouras, 1995; McIlvaine-Newsad *et al.*, 2004).

Conclusion

This study of Fourfold Farm CSA provides an entry point into understanding the central role that culture plays in the alternative paradigm of urban agriculture. Unlike other studies of CSAs that focus on farmers (e.g., Lass *et al.*, 2001; Tegtmeier and Duffy, 2005) or members (e.g., Kolodinsky and Pelch, 1997; DeLind and Ferguson, 1999), this study focuses on the cultural aspects of CSAs, as exemplified by civic engagement, community and the celebration of local food.

In particular, this study highlights the importance of culture in two ways. First, it takes as axiomatic that social-science investigations of cultural aspects of agri-culture are an essential component for building our understanding of all forms of agriculture (urban and rural; conventional and alternative) and for setting a course for improving the relationships among people, land and nature. Our focus on the cultural practices that underscore and support this particular CSA can offer fodder for future investigations.

Second, concentrating on the cultural elements of urban agriculture, as manifested in CSAs and expressed as civic engagement, community and the celebration of local food, enhances our understanding of the alternative paradigm. Indeed, this study puts at the forefront the complex interrelationships between food producer and food consumer – relationships that take place within the overarching context of culture.

In addition, this study provides an entry point into answering Pretty's (2002) questions noted at the outset of this paper. By posing these two questions, he has opened the door to linking sustainability and farming through the medium of culture.

1. *Can we put the culture back into agri-culture without compromising the need to produce enough food?* This study of Fourfold Farm CSA illustrates some of the ways we can put the culture back into agriculture while still producing adequate amounts of healthy food for urban populations. In particular, the study illustrates the extent to which the cultural component helps sustain members' engagement in the farm's operation (thereby supporting its very existence) as well as providing essential links between the farm and the broader community. CSAs, unlike other, more conventional forms of agriculture, exist *because* of the interdependent relationships fostered between farm and community, and many of these relationships are built through the cultural practices of fostering civic engagement, building community and celebrating local food. By focusing on the cultural aspects of the Fourfold Farm CSA, our research illuminates the extent to which these practices are not only important to members but also critical to ensuring the sustainability of the farm itself.

2. *Can we create sustainable systems of farming that are efficient and fair and founded on a detailed understanding of the benefits of agro-ecology and people's capacity to cooperate?* As Pretty (2002) emphasizes, industrial farming, as part of the conventional paradigm, has shorn the culture from agri-culture, turning food into a commodity and farming into what McWilliams (2000) has referred to as 'factories in the field'. If the alternative paradigm is to evolve into sustainable systems of farming, then culture becomes the key to returning food and farming to their rightful places. In effect, culture can contribute to the paradigm shift necessary to move toward sustainability. It can raise consciousness, provide a forum, bestow legitimation and open up the opportunity for farming and sustainability to meet. In doing so, culture creates the climate where efficiency, fairness, agro-ecology and cooperation can emerge and thrive.

As a form of the alternative paradigm, urban agriculture carries enormous cultural potential. This study steps through the door that Pretty (2002) opens to suggest that urban agriculture, in the form of CSAs, can be at the vanguard of putting the culture back into agriculture and creating sustainable systems of farming.

References

Abbott Cone, C., Kakaliouras, A., 1995, 'Community supported agriculture: building moral community or an alternative consumer choice', *Culture and Agriculture* 15 (51–52), 28–31.

Babbie, E., 2001, *The Practice of Social Research* (9th edn), Wadsworth/Thompson Learning, Belmont, CA.
Beus, C. E., Dunlap, R. E., 1990, 'Conventional versus alternative agriculture: the paradigmatic roots of the debate', *Rural Sociology* 55 (4), 590–616.

Bourque, M., 2000, 'Policy options for urban agriculture', in: N. Bakker, M. Dubbeling, S. Gündel, U. Sabel-Koschella, H. de Zeeuw (eds), *Growing Cities, Growing Food: Urban Agriculture on the Policy Agenda*, Deutsche Stiftung für Internationale Entwicklung, Germany, 119–146.

Charmaz, K., 2006, *Constructing Grounded Theory*, Sage, London.

Chiappe, M. B., Flora, C. B., 1998, 'Gendered elements of the alternative agriculture paradigm', *Rural Sociology* 63 (3), 372–394.

Chung, K., Kirkby, R. J., Kendell, C., Beckwith, J. A., 2005, 'Civic agriculture: does public space require public ownership?', *Culture and Agriculture* 27 (2), 99–108.

Cosgrove, D., 2000, 'Culture', in: R. J. Johnston, D. Gregory, G. Pratt, M. Watts (eds), *The Dictionary of Human Geography* (4th edn), Blackwell, Malden, MA, 143–145.

DeLind, L. B., Ferguson, A. E., 1999, 'Is this a women's movement? The relationship of gender to community-supported agriculture in Michigan', *Human Organization* 58 (2), 190–200.

Fieldhouse, P., 1996, 'Community shared agriculture', *Agriculture and Human Values* 13 (3), 43–47.

Henderson, E., 2007, *Sharing the Harvest: A Citizen's Guide to Community Supported Agriculture*, Chelsea Green, White River Junction, VT.

Johnson, A. G., 2000, 'Culture', *The Blackwell Dictionary of Sociology: A User's Guide to Sociological Language* (2nd edn), Blackwell, Oxford/Malden, MA.

Johnston, R., 2000, 'Community', in: R. J. Johnston, D. Gregory, G. Pratt, M. Watts (eds), *The Dictionary of Human Geography* (4th edn), Blackwell, Malden, MA, 101–102.

Kolodinsky, J. M., Pelch, L. L., 1997, 'Factors influencing the decision to join a community supported agriculture (CSA) farm', *Journal of Sustainable Agriculture* 10 (2/3), 129–141.

Lass, D., Bevis, A., Stevenson, G. W., Hendrickson, J., Ruhf, K., 2001, *Community Supported Agriculture Entering the 21st Century: Results from the 2001 National Survey*, University of Massachusetts, Amherst, MA.

Lyson, T. A., 2004, *Civic Agriculture*, Tufts University Press, Medford, MA.

Lyson, T. A., 2005, 'Civic agriculture and community problem solving', *Culture and Agriculture* 27 (2), 92–98.

McDonald, J. H., 2005, 'Keeping culture in agriculture: a call for discussion', *Culture and Agriculture* 27 (2), 71–72.

McIlvaine-Newsad, H., Merrett, C. D., McLaughlin, P., 2004, 'Direct from farm to table: community supported agriculture in Western Illinois', *Culture and Agriculture* 26 (1/2), 149–163.

McIntyre, B. D., Herren, H. R., Wakhungu, J., Watson, R. T. (eds), 2009, *Agriculture at a Crossroads*, International Assessment of Agricultural Knowledge, Science and Technology for Development, Volume IV, North America and Europe, Island Press, Washington, DC.

McWilliams, C., 2000, *Factories in the Field: The Story of Migratory Farm Labor in California*, The University of California Press, Berkeley, CA.

Millstone, E., Lang, T., 2003, *The Penguin Atlas of Food*, Penguin Books, New York.

OED Online, 2009, 'Culture', *Oxford English Dictionary* [available at www.oed.com].

Pretty, J., 2002, *Agri-Culture: Reconnecting People, Land and Nature*, Earthscan, London, Sterling, VA.

Roberts, P., 2008, *The End of Food*, Houghton Mifflin, Boston, MA.

Robyn Van En Centre, 2009 [available at www.csacentre.org].

Smith, D., 2005, *Institutional Ethnography: A Sociology for People*, AltaMira Press, Lanham, MD.

Stake, R., 2005, 'Qualitative case studies', in: N. K. Denzin, Y. S. Lincoln (eds), *The Sage Handbook of Qualitative Research* (3rd edn), Sage, London, 443–446.

Strauss, A., 1987, *Qualitative Analysis for Social Scientists*, Cambridge University Press, Cambridge.

Sumner, J., 2003, 'Visions of sustainability: women organic farmers and rural development', *Canadian Woman Studies* 23 (1), 146–150.

Tegtmeier, E., Duffy, M., 2005, *Community Supported Agriculture (CSA) in the Midwest United States: A Regional Characterization*, Leopold Center for Sustainable Agriculture, Iowa State University, Ames, IA.

Van En, R., 1995, 'Eating for your community', *In Context* 42 [available at www.context.org/ICLIB/IC42/VanEn.htm].

Yin, R. K., 2003, *Case Study Research: Design and Methods* (3rd edn), Applied Social Research Methods Series, Vol. 5. Sage, London.

The emergence of urban agriculture: Sydney, Australia

David Mason[1]* and Ian Knowd[2]

[1] Department of Industry and Investment NSW, Locked Bag 4, Richmond 2753, NSW, Australia
[2] School of Social Sciences, University of Western Sydney, Locked Bag 1797, Penrith South D.C., 1797 NSW, Australia

Across the world the phenomenon of urban agriculture (UA) is defining itself after emerging from a mainly grass-roots response, evidenced in the Sydney Metropolitan Region by the Hawkesbury Harvest phenomenon and the Sydney Food Fairness Alliance, to powerful global forces which are negatively and paradoxically impacting on the quality of life of urban and farming communities. In the developed world these major forces include: (1) urban sprawl and its progressive sterilization of agricultural lands; (2) the supermarket dominance of food chains; (3) the fast food industry and associated health problems such as obesity; (4) globalization. The community-based promotion and marketing of local agriculture is causing some governments and public and private organizations throughout the world to recognize UA as a strategic mechanism to enable urban communities to deal with food security in the context of neo-liberalism, climate change, pandemics, natural disasters, human and environmental health, carbon footprint, biosecurity/terrorism, peak oil, waste management, and landscape and natural resource management. This paper explores the history of UA in the Sydney region. It is a narrative that allows for UA in the Greater Sydney Metropolitan area to draw on the experiences of other developed countries where UA is establishing its position.

Keywords: food systems, planning, sustainability, urbanization, urban agriculture

Introduction

Many cities around the world have their own stories of how and why agriculture as a land use and human activity has been overridden by a range of local, regional and global forces including urbanization, food chain dominance, free market economics and changed eating patterns. This particular narrative is about Sydney in New South Wales (NSW), Australia. It tells of Sydney's potential, what has been lost, and how, through grass-roots community action, agriculture in the urban and urbanizing environment is re-establishing itself as an important aspect of Sydney's culture.

It is a story that has many resonances with similar situations and communities around the world. It is a local instance of a global phenomenon, and the emergence of urban agriculture articulates common themes of development, urbanization and globalizing markets in Sydney and the wider world. There is a salience to this story that makes it worthy of attention, instructive in what it has to teach us, and portentous in its capacity to predict the political and social pressures that agriculture in urbanizing environments imposes on planners and communities.

The vision splendid

Lieutenant Lachlan Macquarie arrived in Port Jackson (Sydney Harbour) on 28 December 1809 to take up his commission as the Fifth Governor of the colony of NSW on 1 January 1810. Prior to his arrival a culture of inequitable land granting and dealing had established itself in Sydney. The position of governor was the supreme authority in the colony for the first 35 years following the arrival of the First Fleet in 1788. Some early governors compromised that supremacy by currying favour with those who had established themselves as influential and well-connected in the Colony, particularly with officers of the army, by granting them

*Corresponding author. Email: david.mason@industry.nsw.gov.au

INTERNATIONAL JOURNAL OF AGRICULTURAL SUSTAINABILITY 8 (1&2) 2010
PAGES 62–71, doi:10.3763/ijas.2009.0474 © 2010 Earthscan. ISSN: 1473-5903 (print), 1747-762X (online). www.earthscan.co.uk/journals/ijas

earthscan

disproportional favours. So much so that when Macquarie's predecessor, Governor William Bligh (1806–1808) challenged the near-monopoly of trade and land grants being exercised by officers of the NSW Corps and their associates amongst the leading landowners, he was arrested by the army. This was Australia's only military coup and for the next two years, until the arrival of Macquarie, officers of the Corps took the role of governor upon themselves. The arrival of Lieutenant Macquarie with his own regiment in 1810 restored the power of the governor and saw the NSW Corps disbanded (Parliament of NSW, 2008).

During the week commencing 29 November 1810 Governor Macquarie, his wife and party travelled by horseback and carriage along the banks of the Hawkesbury-Nepean River from Jamisontown (now a suburb of Penrith) to Green Hills (renamed Windsor). On Thursday 6 December 1810 Governor Macquarie proclaimed the five towns of Wilberforce, Pitt Town, Windsor, Richmond and Castlereagh. The purpose of these towns was to service the surrounding farming areas to ensure the fledgling and rapidly expanding settlement of Sydney, 60km to the southeast, had a reliable food supply.

Soon extensive farming fed and clothed the infant colony, taking advantage of the wide range of fertile soils and microclimates of the Hawkesbury-Nepean River catchment. This proved to be the most successful strategic food security intervention by government in the new colony since the arrival of the First Fleet under the command of Captain Arthur Phillip at Sydney Cove on 26 January 1788. The cargo of human rejects from England, disgorged from the bowels of the prison ships at that time, struggled for the next three decades to feed itself.

Macquarie's journal records (Saturday 1 December 1810):

> Mrs. M. and myself were quite delighted with the beauty of this part of the Country; its great fertility, and its Picturesque appearance; and especially with the well-chosen and remarkable fine scite [sic] and situation of the Government Cottage and Garden on the Green Hills (Macquarie, 1787–1824; Hawkesbury Historical Society, 2009).

Macquarie saw the vision splendid – an agrarian civil society he hoped, despite the paradox of harnessing powerful forces of self-interest for public good, would ensure an egalitarian legacy for future generations of Sydney residents. In the annals of colonial and modern histories of NSW and Australia, he is lauded as a man for the people and the public good.

How has that vision fared?

Following his return to England in 1820 the powerful forces of self-interest which Macquarie had fought against re-established their domination over the development of the colony. This development was essentially unstructured and oriented to quick wealth acquisition. The new Colony was the land of opportunity and there was plenty of land available to exploit; wealthy landowners quickly developed the template of land acquisition and break up. Today there is a significant perception in the Sydney community and indeed right across Australia that such an exploitive and influential culture still flourishes, expressed as developer-driven subdivision.

Spearritt (1999) provides a comprehensive examination of the subdivision boom in Sydney that began in the 1920s to meet the demand by prospective home-owners, wistful thinkers and speculators, fuelled by the real estate industry, with the fingers of urbanization following tram and train infrastructure development. In the 1940s, housing land size in these urbanizing areas had generally settled out to what became known as the 'quarter acre block' and to acquire such was marketed as the 'home owner's dream' for much of the rest of the 20th century. In areas on the fringe of the tram and train infrastructure, rural subdivision was marketed as farmlets (1–10 acre blocks) for high return on produce, be it fruit, vegetables or meat – particularly poultry. In the areas beyond such infrastructure influence, broad area agriculture remained the main form of land use in the Sydney region until the second half of the 20th century (Waterhouse, 2005).

The 1940s marked the period when Sydney's land use began to change significantly. The suburbanization that started to sprawl out across the region, described variously as 'promiscuous urbanization' (County of Cumberland, 1948, p. 129) and a 'rolling wave' (Rutherford et al., 1967, cited in Johnson et al., 1998; Kelleher, 2001), was increasing its pace to indifferently consume the farming land. Subdivision had become embedded into Sydney's collective psyche (Ashton and Freestone, 2008). Immigrants who came to Australia after the end of World War II and acquired or leased land on the outskirts of ever-expanding urban Sydney soon realized the financial bonanza to be had from using the land to grow food and plants in their various forms, while waiting until the housing arrived on their doorstep. When the housing did arrive, many simply sold to the developers and bought further out, and waited for the next housing wave to arrive.

By the 1980s, agricultural land in the Sydney region was regarded politically and within the bureaucracy as 'land awaiting higher economic development' (Gillespie and Mason, 2003, p. 1), a reflection of the thinking of the majority of the landowners in the region. At the Royal Australian Planning Institute Conference at Penrith in Western Sydney in 1993, in response to a question about planning for agriculture in the Sydney region, a planning official replied: 'There is no place for agriculture in the Sydney region. Agriculture belongs over the (Great Dividing) Range and any agricultural land is land awaiting higher economic development.'

Towards the late 1990s, 'rural lifestyle living' had overtaken farming as the major land use of acreage blocks (Sinclair, 2001). Lot numbers per acre in urbanizing areas had increased from four to eight. Planning that accommodated the neo-liberal (economic rationalist) principle of land use being determined by market forces was the prime reason for this situation. There is widespread acceptance that neo-liberalism is expressed through the manifestation of the urban sprawl of our modern cities (Knowd, 2009, citing Cook and Ruming, 2008). Lands used for farming were and still are ideal for housing development. All the clearing and much of the development work has been done by the farmers.

The dangers of Sydney's rural lands being totally overwhelmed by urban development were noted as worthy of attention as far back as the County of Cumberland Plan (1948). This plan recognized the need to balance city and country (County of Cumberland, 1948, p. 124). It introduced a greenbelt which defined how the city would be structured by zoning (land-use categories) and zones (urban/non-urban) (Bunker and Holloway, 2001). This greenbelt was lost when the second metropolitan strategy, the Sydney Region Outline Plan, was released in 1968 because of population growth and the successful lobbying of the housing and development industries (Ashton and Freestone, 2008).

Despite zoning being introduced as the mainstay of planning controls in NSW in the 1940s, there has never been any political foresight or will to put in place the protections that other international jurisdictions have done for agriculture associated with large cities. An outstanding example is the British Columbia Agricultural Land Commission Act of 1972 to protect the Hudson River delta in Vancouver. There are significant similarities between that delta and the Hawkesbury-Nepean catchment and the associated relationships with their respective cities.

All the plans that have followed the Cumberland Plan have failed to accurately describe and acknowledge the relationship between farming and the city of Sydney, and to conceive of the implications of lost agricultural lands for the city's food system and food culture (Knowd, 2006b):

- Metropolitan Strategy 1988 – recognized the need to protect parks, forests and catchments;
- Sydney's Future 1993 – recognized water quality, waste management, bushland and landscape conservation;
- Cities of the 21st Century 1995 – recognized 'prime land' (with no subsequent action);
- Metropolitan Strategy 2004 – Direction 3: Manage Growth and Value Non-Urban Areas – recognition will be given to non-urban land so that it is not treated as 'land in waiting' for urban development (NSW Planning, 2005).

Another factor that has impacted on the realization of Macquarie's vision relates to food equity. Until the 1940s, individual access to fresh local food for Sydney-siders was at its premium. The size of the Sydney region, its range of climatic conditions and soil types ensured there was an abundance of food to meet the needs of the burgeoning settlement. Having overwritten the indigenous form of land use, a relationship with agriculture and its by-products came to be the norm as farming expanded to be the predominant form of human activity in the Sydney region.

The food chain was essentially producer driven. Many people grew their own vegetables and had their own fruit trees on their farms or in their backyards. Those who did not had access to supplies through other mechanisms, such as that which evolved into the corner store. Dairies within easy access on the edge of the settled areas produced milk, cheese and butter requirements, spreading a final layer onto what many came to take for granted as Sydney's food security.

The emergence of the supermarket system in the food chain in the second half of the 20th century has predisposed us to the loss of direct connection between people and farmers and the food they produced as well as the demise of the corner grocery store. Relationship-based social and environmental benefits were being progressively traded off for convenience shopping. By the end of the 1990s the total dominance of the food chain by the major supermarkets was impacting on farm economic performance. Farm holdings in the Sydney region, which had been getting progressively smaller due to the cost of land being pushed up by the urbanizing situation, were under extreme pressure from increased competition and reduced power to determine price and thus incomes (Knowd, 2006a). The increasing pace of life and associated distractions of modern living were diminishing people's interest or capacity to grow their own food or indeed how to cook a balanced meal. These factors also contributed to the rise of a third force – the fast food industry.

The combined effect of all of this on the wider community was that community health interests were fighting a losing battle against the effects of changes in the food system. These effects are believed to play a role in the disturbing increases in lifestyle diseases such as obesity and the increasing exclusion of the more disadvantaged within the community from access to fresh local food (Saville, 2002). It is indeed ironic that when the first Sydneysiders had easy access to an abundant supply and variety of local fresh food, they did not have the associated knowledge of its importance to human health. Today we face the reverse situation. The Sydney region's once cornucopic food system is under threat at the very time when awareness of linking health and well-being to fresh local food is rising sharply, and consumer demand for local foods is gaining momentum.

Is Macquarie's vision of an agrarian (land and food) based equitable society for Sydney a lost cause? One could be excused for thinking so. However, there may be hope due to some global, regional and local forces contributing to the emergence of public attention on the issue of local and regional agriculture in the Sydney region.

The emergence of public attention on the promotion and marketing of local food

There is experiential evidence in the Sydney region to support the proposal that the possibility of achieving land and food equity and associated security of both in the region is enhanced by the emergence of several global, regional and local forces since the 1990s. A fundamental principle common to these forces is the strategic linkage of local and regional agriculture to other policy arenas such as human health, environmental health, tourism, food security, food safety, bio-security, climate change and landscape management.

At the global level the Agenda 21 'Healthy Cities' programme that emerged from the Environment and Development Conference in Rio de Janeiro in 1992 provided ratification of the establishment of the Penrith Food Project in 1991 (Webb *et al.*, 2001). The founding

partners of the project were the Wentworth Area Health Service, Penrith City Council, Penrith Food Policy Committee and the NSW Department of Community Medicine (Westmead Hospital). In 1993 the NSW State Government Department of Agriculture became a member of the Project and remained so throughout the decade.

The project was a multi-strategy approach to address many of the institutional and community infrastructure problems that prevented people from obtaining and choosing a healthy diet in the local area (Reay and Webb, 1998). During this time the concept of 'open farm days' was introduced where the Sydney community was invited to visit local Penrith farms to meet the growers and buy their produce direct. This proved to be very successful.

The Penrith Food Project provided the model for the establishment of the Hawkesbury Food Program in the late 1990s. A characteristic of the programme was the reorientation of food supply systems consistent with the Ottawa Charter for Health Promotion, Agenda 21, and the Hawkesbury City Council Healthy Cities initiative 'Growing Shared Solutions to a Healthy Hawkesbury Community' (Saville, 2002).

Hawkesbury Food Program Partners and Steering Committee members included the Hawkesbury District Health Service, Wentworth Area Health Service, Hawkesbury City Council, NSW Agriculture, University of Western Sydney – Hawkesbury Campus, Hawkesbury Skills and Earthcare, Food for All and local community agencies. It was from the Hawkesbury Cuisine subcommittee of the programme that the idea of linking local agriculture with tourism was generated in early 2000. This quickly led to the establishment of Hawkesbury Harvest with its very successful agritourism arm, the Farm Gate Trail. The interests represented by those who came together to create Hawkesbury Harvest included health, agriculture, tourism, hospitality and education.

The Hawkesbury local government area was ideal for this grass-roots, community-based innovation. During the 1990s and the first decade of the new millennium, the council had conducted a number of community surveys to determine, among other things, why people chose to live in the Hawkesbury local government area. The overwhelming response was because of the rural amenity (HCC, 2000, 2004). Hawkesbury Harvest has provided the instrument to educate the Hawkesbury constituents about the relationship between agriculture and rural amenity. The organization has played a leading role in putting Sydneysiders in touch with the agriculture of the Sydney region. Its four Farmers' and Fine Food Markets, Farm Gate Trail, open farm days, special events, Slow Food Convivium and provedore service have attracted a great deal of media attention. The original 13 Farm Gate destinations on the Farm Gate Trail in 2000 in the Hawkesbury local government area has expanded to nearly 100 across five local government areas around the perimeter of Sydney. Tens of thousands of people visit the Farm Gate Trail each year and up to 100,000 people listen to the five-minute Farm Gate Trail segment every Saturday morning on Sydney's ABC radio.

There are an increasing number of players in the emerging local and regional food culture of Sydney. The Farmers' Market movement has brought more and more Sydneysiders into contact with local and regional farmers. There are more than 20 farmers' markets in the Sydney region. The movement has created such interest it seems that everyone wants a farmers' market. Such demand and the limited number of growers able or prepared to sell through the farmers' market system has raised the issue of authenticity (Graham, 2009).

A hybrid community/government grass-roots player is the Sydney Food Fairness Alliance. This organization has arisen from community health interests in food equity for those Sydneysiders who do not have access to or enjoy the health benefits of fresh local food. It was created in 2005 by a group of health workers, nutritionists, community garden advocates and permaculturists. The role of the Alliance includes advocacy, education and lobbying for affordable food and capacity to influence the NSW State Government's Metropolitan Strategy (Knowd, 2009). This group offers a great deal to the potential of Macquarie's vision of food equity being realized.

What has been the government response to this increasing community interest in local food? The answer to that question has an historical context. In 1993 NSW Agriculture, the State's government agriculture agency provided a presence in Sydney specifically to deal with the issue of agriculture in the city's urban and urbanizing situation at a strategic level. An outbreak of blue-green algae in the Hawkesbury-Nepean River was the trigger for the intervention. It was soon realized the issue of agriculture in the region could not be dealt with purely on an environmental basis.

NSW Agriculture began a process of consultation with social, industry, environment, and local and state government organizations and agencies on the issue, leading to the formation of a committee to develop a strategic plan for agriculture. The process was facilitated by NSW Agriculture, resulting in the release of the *Strategic Plan for Sustainable Agriculture – Sydney Region* by the then Minister for Agriculture, The Hon. Richard Amery in May 1998 (NSW Agriculture, 1998). Minister Amery's electorate was in Western Sydney and he recognized the value of agriculture to Sydney. That recognition was substantiated in 2003 when it was established that the farm gate value of agriculture in the Sydney region was $1 billion (AUD). This represented 12 per cent of the State's total production on less than one per cent of the state's agricultural lands (Gillespie and Mason, 2003).

The strategic plan, which was the first formal strategic intervention by government in the Sydney region since 1810 in regard to the security of agriculture, was not prescriptive but rather provided a framework for sustainability with the document withstanding the test of time. The innovation it articulated has diffused upwards and downwards in state and local government bureaucracy since its release, being a force for change at a regional level.

Having provided an instrument with which to engage the bureaucracy in the issue of sustainable agriculture in the Sydney region, NSW Agriculture, along with the University of Western Sydney-Hawkesbury Campus, facilitated the strategic development of Hawkesbury Harvest. It was recognized that if agriculture was to have a sustainable role and place in the Sydney region it had to have community and political support. Hawkesbury Harvest presented an ideal mechanism to do that. NSW Agriculture and the university have been members of the Hawkesbury Harvest board since inception. The seeds of Hawkesbury Harvest were sown by Governor Macquarie when he first visited the farms on the Hawkesbury-Nepean flood plains. The similarities between the ground he rode over with Mrs Macquarie back in 1810 and the points of destination on the first Farm Gate Trail map in 2000 are remarkable. Macquarie's vision is Hawkesbury Harvest's vision.

Hawkesbury Harvest has been the major mechanism by which the local, state and federal levels of government have unwittingly made some contribution to the realization of the Macquarie vision. This has been essentially through providing public funding to develop and expand the organization's activities around Sydney's perimeter. Hawkesbury Harvest also engaged with Lend Lease and the General Property Trust as a partner in their successful tender to develop the Rouse Hill Town Centre, a 120ha retail, commercial, housing and environmental complex in northwest Sydney. The development provides a designated 'market square' where Hawkesbury Harvest conducts one of its farmers' and fine food markets.

In 2004 the Department of Infrastructure, Planning and Natural Resources released its Western Parklands Management Vision Summary Report. The NSW Government has been conserving open space corridors for recreation and the environment as part of a long-term plan in Western Sydney for more than 30 years. This 5500ha of land stretches for 26km from Blacktown to Liverpool along Eastern Creek and the hills of Hoxton Park. The parklands will make a major long-term contribution to restoring the balance between nature and growth in Sydney, and enable Western Sydney's community to connect with the outdoors while living in Australia's largest city. It will unlock the potential of this vital public asset to make a fundamental contribution to greening Sydney, for the future of Western Sydney's diverse communities. Provision is made for a sustainable agriculture precinct of more than 500ha within the parkland (Department of Infrastructure, Planning and Natural Resources, 2004).

In December 2008, the NSW Department of Primary Industries (formerly NSW Agriculture), with assistance from Penrith City Council and the NSW Department of Planning, hosted a forum: *Sydney's Agriculture – Planning for the Future*. This forum provided an opportunity for agricultural specialists from across industry, planning, business and government sectors to:

- Examine key issues affecting the future of agricultural production in the Sydney region in the context of projected population growth, continued urban development and loss of productive agricultural land;
- Explore strategies for ensuring the sustainability of agriculture in Sydney to secure the supply of fresh food to the city's growing population (Elton Consulting, 2009).

This forum was called for by the NSW Minister for Primary Industries, the Hon. Ian Macdonald, in response to representations from community, industry and local government seeking his support to retain agriculture in the Sydney region. Approximately 120 people attended the forum. An outcome of the forum was the formation of the Sydney Agriculture Reference Group to make recommendations to broaden and inform the land-use planning, which is meeting on a regular basis to formulate its recommendations.

The formation of this reference group is a physical expression of the demand by community, industry and local government for the major strategy and policy changes, particularly around planning, set out in the *Strategic Plan for Sustainable Agriculture – Sydney Region*, to take place. It demonstrates how long the demand for change can take to develop its own physical force. The next phase is for change actually to begin to happen on-ground. Change takes time.

Consider therefore how long before any change may occur, if at all, as a result of the Rural Industries Research and Development Corporation study titled *Sustainable Small-scale Agri-industrial Projects Neighbouring the Greater Blue Mountains* (2006). The study was conducted in the Bilpin district on the northwest perimeter of Sydney. The purpose of this project is to facilitate the long-term sustainability of agri-industries along the northeastern boundary of the Greater Blue Mountains World Heritage Area, to enhance the buffer between protected conservation areas and the Sydney catchment. The land at this point is in transition assisted by local real estate agents promoting agricultural land in the area for sale as 'land banks' (Knowd, 2006a, 2009). This is occurring despite the Sydney Metropolitan Strategy specifically stating that rural lands will not be regarded as land banks awaiting development (Direction 3: Manage Growth and Value Non-Urban Lands).

The 2010 Macquarie Bicentenary provides an opportunity

The year 2010 marks the bicentenary of the arrival of Governor Lachlan Macquarie and his wife Elizabeth in 1810. This event will be celebrated at many levels in the Sydney community. It provides the three levels of government in Australia with the unique opportunity to lay down the foundations to realize the Macquarie Vision in a postcolonial context – an egalitarian food and land system where the public good is afforded balance and equity with the private good and self-interest of the housing, development and food retailing industries. This would be a first for Australia and would provide a model for other Australian and international cities.

What would be required to achieve this? The answer lies in looking at what is working successfully for agriculture in the urban and urbanizing environments of other cities throughout the world.

International concepts and trends for sustainable UA in the Sydney Basin

Innovation and adaptability of agriculture in the urbanizing environment

The Dutch survive because of innovation. If they did not, much of The Netherlands would be under water. They are also very pragmatic and, in regard to their agriculture, they view Holland as a 'suburb' of Europe that specializes in meeting niche markets throughout the world. They have a fundamental belief that agriculture has the capacity to adjust to urbanization and in so doing create a knowledge industry. The Dutch focus is largely on meeting the niche demand for producing food for health and fashion, and plants generally for fashion (Smeets, 2006).

The Dutch have been able to retain aspects of their traditional agriculture such as larger area dairying (due to community demands) and provide for hi-tech agriculture, as evidenced with its glasshouse industry. A mechanism through which significant inroads are being made in UA is an organization called Innovatie Netwerk. It was established in 2000 by the Dutch Government at the request of various parties in the community. Its purpose is to facilitate system-based innovation to optimize the outcomes from developing sustainable relationships between rural areas, agriculture and the health status of the Dutch people through nutrition.

Innovatie Netwerk operates outside, but in association with, academia, government, bureaucracy and industry. It actively seeks and encourages champions within those sectors as well as at the community, environment, farmer and political levels. It has a database of approximately 5000 people, of whom 200 could be working on one or more of its 32 projects at any one time. It is recognized and accepted that not all projects will succeed and that failure is part of the cost of innovation.

One of the 32 projects it has initiated is the New Villages concept (a village can be greater than 16,000 people). This project includes various projects for identifying possibilities for combining some seemingly incompatible wishes in rural areas. It is not just a question of creating possibilities for living in the countryside, but also about improving rural quality, recognizing regional differences, taking into account future needs for water and storage and meeting future residents' social and cultural needs (Mason, 2006).

What also needs to be recognized is the adaptive capacity of agriculture to urbanization. An organization which is playing a part in the adaption of Dutch agriculture is Educational Training Consultants (ETC International Group). The work of ETC embraces both developed and Third World countries – it is systemic in terms of disciplines as well as spatially on a local, regional, national and international scale. Its work is focused on:

- facilitating capacity building for farmers;
- tapping into and utilizing farming knowledge in urban and urbanizing environments;

- reforming the functions of agriculture and rural areas as an adaption to urbanization;
- influencing agricultural policy through a participatory and multi-stakeholder approach.

The Netherlands and many parts of Europe are urbanizing at a rapid rate. At the same time there is a transition of agriculture from the post-WWII modernization era to the new era of rural development in which agriculture is seen and dealt with as part of a mix of disciplines and stakeholders in the rural environment. The family farm where off-farm income brings urban capital into rural areas, agritourism, direct marketing activities such as farmers' markets, natural resource and landscape management, regional identity through regional branding of regional specific produce and products, organic farming and 'care' farming that caters for intellectually handicapped people are just some of that mix. The Europeans refer to this transition as multi-functionality (Mason, 2006).

Innovation is not just limited to technology and multi-functionality. It also applies to process. In 2004 the Urban Affairs Centre of the University of Toledo and the Centre for Regional Development at Bowling Green State University, Ohio initiated a project designed to optimize the sustainability of the floriculture industry in northwest Ohio (University of Toledo Urban Affairs Centre, 2009). The core of the project is a systemic intervention process that engenders ownership and commitment to the vision, mission and objectives that are determined by the target audience.

The underlying leadership principle employed by the universities is one of 'leading from behind', at the same time empowering those involved. The group that was formed is The Maumee Valley Growers. The role of the universities is a facilitative one. The intervention effectiveness is significantly enhanced by a structure that provides for a grower who is elected to champion the catalyst for change and a 13-member advisory board including representatives from eight greenhouses, Congress Representative Marcy Kaptur's office and Regional Growth Office. Maumee Valley Growers is organized around the concept of an industrial cluster.

Cluster-based economic development is based around the concept that a geographic region, and the businesses contained therein, can compete more effectively when everyone in the region works together for the common benefit of all the stakeholders. Clusters, by definition, include not only business entities such as companies, but also trade associations, financial institutions, vocational training providers, universities, economic development agencies, and any other entity that can be considered important to competition (Mason, 2006).

Local food and direct marketing

There is a distinct trend in the northern hemisphere towards local food and associated direct marketing. In Europe the consumer is having a marked influence on the food chain. A significant factor driving this trend is the food scares that have occurred such as that associated with Chernobyl, mad cow disease outbreaks, foot and mouth disease and the salmonella outbreaks such as that which occurred in the US vegetable industry with resultant deaths.

Research in Canada has established that the trend to local food is on the lower but upward end of the trend curve. It is a bottom-up social values-based trend in food choices rather than one imposed by government or the corporate sector. According to the majority of Canadians the benefits of buying locally grown fruit and vegetables are that they help the local economy (71 per cent), support family farmers (70 per cent), taste better (53 per cent), are cheaper (50 per cent), healthier (44 per cent), safer (44 per cent), environmentally friendly (43 per cent) and preserve green belts (41 per cent) (Babcock, 2006).

The increasing strength of this trend is due to a number of factors, including a consistent and cohesive effort in getting the message out into the public arena. Labelling and branding plays an important role in achieving this (Babcock, 2006).

Research by the New York State Department of Agriculture Marketing (2002) established the increasing role of direct marketing as a mechanism used by local farmers to increase their viability and sustainability as evidenced in Table 1.

Such trends indicate that the demand for local and regional food associated with cities and towns by consumers is slowly but surely moving into the political arena as an issue requiring attention. Just two recent examples at the local level are Sydney Food Fairness Alliance's Food Summit and the Hawkesbury Foundation and University of Western Sydney Urban Research Centre's Feeding Sydney conference, specifically addressing urban planning and food systems. In the US context there is Michael Pollan's (2008) feature piece in *Farmer in Chief*, an open letter to President-Elect Obama in *The New York Times*, 9 October 2008. These social and political trends are replicated in both the developed and underdeveloped world.

Table 1 | **The increasing role of direct marketing by farmers**

Direct marketing activity	Farmers 1987	Farmers 2000	% change
Roadside	859	1043	+12.4
Open stand	1002	1622	+61.9
Pick your own	748	1475	+94.8
Farmers' market	712	1690	+137.4
Other	1130	1796	+58.9
Number of farmers	6125	6667	+8.8
Sales	$112.5 m	$230.2 m	+105

Planning

Planning is the major factor that will determine whether UA will have a place in Sydney's future. It is the issue accorded the highest priority in the *Strategic Plan for Sustainable Agriculture – Sydney Region*.

Sydney's land had been so irreversibly fragmented, there is no way the rigid application of a definitive divide between urban and country could be applied, as is the case in Britain and many other parts of Europe. However, there are areas of land that are suitable for the various forms of UA. The UA spectrum extends from the social and community forms such as community gardens at one end to the profit oriented, hi-tech forms at the other, such as the mushroom industry. In between these is a range including cottage industries, niche production, and market gardens (Mason and Docking, 2005). A land-suitability map to cater for the various forms needs to be completed as a first step.

The argument often used to justify the spread of urban growth is that Sydney has a requirement to accommodate another one million people by 2025. By contrast, the City of Vancouver has been able to accommodate an additional one million people in the region over 30 years (1971–2001) while maintaining productive farmland, important greenspace and habitat. In the 1970s, the British Columbia Government responded to the increasing loss of agricultural land to development by establishing the Agricultural Land Reserve (ARL) administered by the Agricultural Land Commission. The commission has the responsibility for protecting British Columbia's agricultural land. It is supported by the Agricultural Land Commission Act 1972, which takes precedence over, but does not replace, local government regulations and policies.

In the City of Vancouver, the ARL has become the cornerstone of a comprehensive planning approach as embodied in the Liveable Region Strategic Plan of the Greater Vancouver Regional District. These mechanisms are complemented by the Greater Vancouver Region Sustainability Dialogues that establish an institutional context and forum for ongoing consultation on a range of issues including agriculture (Metro Vancouver, 2009). This in particular demonstrates the political will regarding the sustainability of agriculture in the Vancouver Region. Over 15 years (1986–2001), the Greater Vancouver Regional District realized a $400 million increase in its total gross farm receipts, and the area of the ARL remained relatively constant. The annual average net loss of agricultural land fell from a high of 405ha during the 1979–1983 period to a low of 85ha during the 1994–1998 period (Smith and Haid, 2004).

Compare this to the Sydney region where in the four years to 2000 there had been a 10.5 per cent (8600ha) reduction in the land available to agriculture in the Sydney region to 76,900ha (using Australian Bureau of Statistics figures as cited in Gillespie, 2003). At the same time there was a three per cent increase in the number of agricultural holdings in the region. The result is that less available land is being subdivided into an increased number of smaller holdings, and a greater intensity of production (Gillespie, 2003).

In Ontario, Canada, the Places to Grow Act (2005) enables the government to designate growth plan areas and develop growth plans. The Growth Plan for the Greater Golden Horseshoe is the first growth plan to be approved under the Act. The Greater Golden Horseshoe Growth Plan aims to meet the projected increase of population by 3.7 million to 11.5 million between 2001 and 2031. It also provides for the 1.8 million acre greenbelt that protects environmentally sensitive natural areas and agricultural lands. The Golden Horseshoe fits around the western end of Lake Ontario. The edges of the horseshoe are mostly moraine (a ridge of glacial deposits); however, some is defined by the shore of the lake. Within the climatic shadow of the moraine is an environment that supports flora and fauna unique to the area.

The community became very active in seeking to have that uniqueness preserved. The issue became political and was instrumental in the designation of the greenbelt. Agriculture is preserved within the greenbelt by virtue of its relationship with the environment the community wished to preserve. In this context, the greatest value of agriculture is as a buffer between urban development and the environment. However, it is also valued because of the other social, economic and environmental benefits it provides to the greater community.

Conclusion

There is a challenge at local, state and federal levels of government for the role of agriculture to be recognized and catered for in the Sydney region, so that the agrarian heritages of the region (variety of foodstuffs, cultural contributions and recreational opportunity) can fulfil their potential to contribute to the quality of Sydney's social, economic and environmental life. This requires a sustainable balance between agri-industries and the demands for housing a growing population.

This is easy to say but much harder to achieve in the face of political forces and competing interests over land use in the urbanizing zones around our major cities. What we can see is a growing interest in 're-locating' the producers in our food systems and 're-connecting' them to consumers, in shortening the food chain, not just in terms of food miles (Pretty et al., 2005), but also in terms of more direct relationships with the people and places where our food is produced (Pretty, 2001, 2002). Consumers are also seeking the quality-of-life benefits that growing and consuming their own foods can realize for urbanites (Burros, 2009). Together, these health and well-being dimensions of UA are assuming greater importance in the dialogue about food in our major cities, dialogues that invariably place re-engagement with agriculture through our urban food systems as the top priority.

UA now becomes one of the principal components of sustainable cities, not only feeding the citizens, but also maintaining the strategic virtuous circles of attractiveness that city regions increasingly seek to achieve in a globalizing world. These virtues are in food diversity and quality, food culture, the landscapes of agriculture that create rural amenity and recreational assets, the agri-artefacts of history, and the cultural heritages in present-day production. In the Sydney context this cultural heritage includes the role that immigrant farmers have played in diversifying the food offerings of the city.

In this framework, sustainable cities depend on agriculture, the industrialized system of production, and agri-culture, the alternative cultural forms that we see emerging in our suburbs and driving agri-tourism. If we can achieve this we will realize Macquarie's vision and it will be more splendid than he imagined 200 years ago.

References

Ashton, P., Freestone, R., 2008, 'Town planning', *Sydney Journal* 1 (2), June. Dictionary of Sydney project (www. dictionaryofsydney.org) [available at http://epress.lib.uts.edu. au/ojs/index.php/sydney_journal/index].

Babcock, G., 2006, Canadians see many benefits of locally grown food. *Ipsos Reid* [available at www.ipsos-na.com/news/ pressrelease.cfm?id=3298].

Bunker, R., Holloway, D., 2001, Fringe city and contested countryside: population trends and policy developments around Sydney, *Urban Frontiers Program* 6, University of Western Sydney, Sydney.

Burros, M., 2009, Urban farming, a bit closer to the sun. *The New York Times*, 17 June [available at www.nytimes.com/ 2009/06/17/dining/17roof.html].

County of Cumberland, 1948, *County of Cumberland Planning Scheme Report*, presented to the Hon. J. J. Cahill, Minister for Local Government, 27 July.

Department of Infrastructure, Planning and Natural Resources, 2004, *Western Sydney Parklands Management Vision – Summary Report*, DIPNR, Sydney.

Elton Consulting, 2009, *Sydney's Agriculture – Planning for the Future Forum Outcomes Report*, NSW Department of Primary Industries, Sydney.

Gillespie, P. D., 2003, *Agricultural Trends in the Sydney Region. 1996–2001 Census Comparisons: March 2003*, NSW Agriculture, Environment Planning & Management Sub-Program, Sydney.

Gillespie, P. D., Mason, D., 2003, *The Value of Agriculture in the Sydney Region: February 2003*, NSW Agriculture, Environmental Planning & Management Sub-Program, Sydney.

Graham, T., 2009, To market . . . but where are all the farmers?, *Sydney Morning Herald*, 27 June.

Hawkesbury Historical Society, 2009, *Lachlan Macquarie's Travels along the Nepean Hawkesbury Rivers and his Naming of the Five Macquarie Towns* [available at www. hawkesburyhistory.org.au/articles/macquarie_town.html].

HCC (Hawkesbury City Council), 2000, *Hawkesbury Social Plan: Hawkesbury Community Survey*, Hawkesbury City Council, Windsor, Australia.

HCC (Hawkesbury City Council), 2004, *Draft Strategic Plan 2004–2005*, Hawkesbury City Council, Windsor, Australia.

Johnson, N. L., Kelleher, F., Chant, J. J., 1998, 'The future of agriculture in the peri-urban fringe of Sydney', *Proceedings of the 9th Australian Agronomy Conference*, Wagga Wagga, Australia [available at www.regional.org.au/au/asa/1998/7/ 069johnson.htm].

Kelleher, F., 2001, 'Urban encroachment and the loss of prime agricultural land', *Proceedings of the 10th Australian Agronomy Conference*, Hobart, Australia, accessed from The Regional Institute Limited [available at www.regional.org.au/asa/2001/3/ a/kelleher2.htm].

Knowd, I., 2006a, 'Tourism as mechanism for farm survival', *Journal of Sustainable Tourism* 14 (1), 24–42.

Knowd, I., 2006b, The urbanization of Sydney, Presentation to *Your Land, Your Lifestyle: Our Hawkesbury*, Community Land Use Forum convened by University of Western Sydney, Hawkesbury Rainforest Network, Blue Mountains World Heritage Institute, Hawkesbury City Council, Hawkesbury Nepean Landcare, Hawkesbury Nepean Catchment Management Authority, and Hawkesbury Harvest, 27 August, HAC Theatre, University of Western Sydney, Richmond.

Knowd, I., 2009, '*Hawkesbury harvest: panacea and paradox in Sydney's rural tourism hinterland*', Unpublished PhD thesis, University of Western Sydney, Sydney.

Macquarie, L., 1787–1824, *Journals of Tours in New South Wales and Van Diemen's Land, 1810–1822.* A777–A786, Safe 1/365–

Safe 1/374 (Photocopy: CYA777–CYA786), State Library of New South Wales, Sydney.

Mason, D., 2006, Urban agriculture – to identify how sustainable urban agriculture can benefit the quality of life of Australian communities, *Report to the Winston Churchill Memorial Trust of Australia*.

Mason, D., Docking, A., 2005, *Agriculture in Urbanising Landscapes – A Creative Planning Opportunity*, Planning Institute of Australia Congress, Melbourne.

Metro Vancouver, 2009, *Sustainable Region Initiative – Board in Brief* [available at www.coquitlam.ca/NR/rdonlyres/ E02D5A2C-A17D-4C96-AAA8-440B1C3A9295/93011/CITY DOCS785674v1RC_Regular_AgendaB_june012009_22.PDF].

New York State Department of Agriculture and Marketing, 2002, *New York Direct Marketing Survey*, New York Agricultural Statistics Service, New York.

NSW Agriculture, 1998, *Strategic Plan for Sustainable Agriculture – Sydney Region*, NSW Agriculture, Orange, NSW [available at www.dpi.nsw.gov.au/agriculture/resources/land/ planning/sydney].

NSW Planning, 2005, *City of Cities: A Plan for Sydney's Future: Metropolitan Strategy*, NSW Department of Planning, Sydney [available at www.metrostrategy.nsw.gov.au/dev/uploads/ paper/introduction/index.html].

Parliament of NSW, 2008, *The Governor of NSW* [available at www.parliament.nsw.gov.au/prod/web/common.nsf/key/ resourcesSystemTheGovernorofNewSouthWales].

Pollan, M., 2008, The food issue – an open letter to the next Farmer in Chief, *The New York Times Magazine* [available at www.nytimes.com/2008/10/12/magazine/12policy-t. html?_r=3&ref=magazine].

Pretty, J., 2001, *Some Benefits and Drawbacks of Local Food Systems: Briefing Note for TVU/Sustain AgriFood Network*, 2 November [available at www.sustainweb.org/pdf/ afn_m1_p2.pdf].

Pretty, J., 2002, *Agri-Culture: Reconnecting People, Land and Nature*, Earthscan, London.

Pretty, J. N., Ball, A. S., Lang, T., Morison, J. I. L., 2005, 'Farm costs and food miles: an assessment of the full cost of the UK Weekly Food Basket', *Food Policy* 30 (1), 1–20.

Reay, L., Webb, K., 1998, *Penrith Food Project, Triennial Report*, Penrith City Council, Penrith.

Rutherford, J., Logan, M. I., Missen, G. J., 1967, *New Viewpoints in Economic Geography*, Martindale Press, Sydney.

Saville, L., 2002, Hawkesbury Food Program. *Sustaining Our Communities, International Local Agenda 21 Conference*, Adelaide, 3–6 March. Environment Australia [available at www.adelaide.sa.gov.au/soc/pdf/saville.pdf].

Sinclair, I., 2001, 'Lifestyle living', in: *A View from the Edge: Issues in Rural and Metropolitan Fringe Planning*. New Planner 49. Planning Institute of Australia, NSW Division, Sydney.

Smeets, P., 2006, Alterra, Wageningen UR, Personal conversation. Churchill Study Tour.

Smith, B., Haid, S., 2004, 'The rural–urban connection: growing together in Vancouver', *Plan Spring/Printemps 2004* 36–39.

Spearritt, P., 1999, *Sydney's Century: A History*, UNSW Press, Sydney, 47–52.

University of Toledo Urban Affairs Centre, 2009, *Northwest Ohio Economic Research Collaboration* [available at http://uac. utoledo.edu/nwoerc/nwoerc.htm].

Waterhouse, R., 2005, *The Vision Splendid: A Social and Cultural History of Rural Australia*, Curtin University Books, Western Australia.

Webb, K., Hawe, P., Noort, M., 2001, 'Collaborative intersectoral approaches to nutrition in a community on the urban fringe', *Health, Education & Behaviour* 28 (3), 306–319.

The challenges to urban agriculture in the Sydney basin and lower Blue Mountains region of Australia

J. Merson[1]*, R. Attwater[2], P. Ampt[1], H. Wildman[3] and R. Chapple[4]

[1] Institute of Environmental Studies, Vallentine Annexe, University of New South Wales, Sydney, 2052 NSW
[2] University of Western Sydney, Richmond, 2753 NSW
[3] Microbial Management Systems, Australia
[4] Blue Mountains World Heritage Institute, Katoomba, 2780 NSW

The western edge of the Sydney basin in Australia has been one of the major sources of fruit and vegetables for the Sydney markets. A rapid expansion of urban development in this region has caused a significant reduction in the number of small farms and market gardens. Urban and peri-urban agriculture in the region also provides an important buffer between urban development and the neighbouring Greater Blue Mountains World Heritage Area. The decline in urban agriculture can be attributed to various causes including urban expansion and economies of scale. This paper presents an overview of a four-year project that explored options for supporting these vulnerable farming communities both in terms of the regulatory factors and economic and environmental sustainability. The role of agri-industries as landscape buffers to the neighbouring World Heritage Area was investigated in relation to resilience, communities of practice, and ecosystem services. The study developed tools, in conjunction with targeted representative landholders, that can assist in enhancing the economic and environmental resilience of these agri-industries. These tools included an integrated bio-system approach to waste using organic waste conversion, and the use of landscape function analysis to monitor across farms to help address erosion, loss of nutrients and inefficient water management.

Keywords: integrated bio-systems, landscape function analysis, protected area buffer zones, urban agriculture

Introduction

Over the past two decades there has been a rapid expansion of urban development at the western edge of the Sydney basin and lower Blue Mountains in New South Wales (NSW), Australia. This region, known as the Hawkesbury Nepean, is a picturesque landscape of orchards and small farms and has been one of the major sources of fruit and vegetables for the Sydney markets. Urban expansion has caused a significant reduction in the number of small farms and market gardens in this region. While this decline in peri-urban agriculture is assumed to be due to the urban expansion, the reality is more complex (Dang and Malcolm, 2006). Market factors have also had a great deal to do with this decline, with cheaper imports from overseas and large supermarket chains sourcing products from regions across the country where farmers can produce at much larger economies of scale. As a consequence, profit margins and capital investments are in decline. There is also the demographic reality of the farmers in the region, with most farmers in their late fifties and whose only asset is farmland. Even where there are children willing to take over the farm, there is often not enough return for two families to make a living under present circumstances. At the same time, land values have risen and this has reduced the capacity to raise the capital to buy farmland, while making the option of subdivision more attractive. The story is a familiar one, and is repeated in many regions of the world. The economically rational response to this situation is to say that this is a natural course of events as markets change and farms make way for urban expansion as the value of land rises ... end of story.

*Corresponding author. Email: j.merson@unsw.edu.au

INTERNATIONAL JOURNAL OF AGRICULTURAL SUSTAINABILITY 8 (1&2) 2010

PAGES 72–85, doi:10.3763/ijas.2009.0464 © 2010 Earthscan. ISSN: 1473-5903 (print), 1747-762X (online). www.earthscan.co.uk/journals/ijas

earthscan

However, the issue is still more complex, for there are other considerations that need to be taken into account when reflecting on the demise of peri-urban agriculture in this area. The Sydney basin region, as its name suggests, is a semicircular coastal plain bounded by National Parks to the north and south and the rugged mountains of the Greater Blue Mountains World Heritage Area (GBMWHA) to the west (see Figure 1). This densely forested region is dominated by eucalypts. In fact, the reason for the region gaining World Heritage status is that it is home to over a hundred of the known eucalypt species. This ecosystem is also fire-dependent and bushfires are a major threat to urban infrastructure in the area – as is characteristic of ecosystems in the Mediterranean and the southwest United States (Bradstock *et al.*, 2002). Further, in the past decade there has been an increase in fire frequency and intensity across the southeast of the Australian continent caused by the hotter and drier conditions that have been experienced as a consequence of climate change (Merson, 2006). The intensity and ferocity of fires in Canberra in 2003, NSW in 2004 and Victoria in 2008/2009 has led to the re-emergence of the debate over the value of buffer zones between urban development and fire-prone and densely forested areas. A similar debate is occurring in southern France and other fire-prone areas of Europe in the face of similar increases in fire frequency and intensity. At present the areas of urban agriculture which border the World Heritage Area serve as a buffer zone, but without a clear regulatory framework this buffer could easily be eroded, making urban and suburban communities more vulnerable to the impacts of bushfire, especially under more extreme conditions.

There are other reasons for maintaining peri-urban agriculture in the region, which include a range of ecosystem services, food production closer to major centres of population in a more carbon constrained world, and meeting growing consumer demand for a more diverse range of local seasonal products linked to the rapidly expanding phenomenon of growers' markets.

Against this background, a four-year research project funded by the Rural Industries Research and Development Corporation (RIRDC) was undertaken to explore the issues confronting these vulnerable farming communities in terms of land-use regulations and their longer term economic and environmental sustainability. This paper presents an overview of the findings of that report (Merson *et al.*, 2009).

Aims

The role of agri-industries as landscape buffers to the neighbouring World Heritage Area was investigated in relation to resilience, communities of practice and ecosystem services. The specific aims were:

- to provide a realistic assessment of the complex social, economic and environmental drivers impacting on the small-scale rural communities in the region, and to represent the diverse modes of production they are involved in;
- to develop a number of tools in conjunction with targeted representative landholders, which could assist in enhancing the economic and environmental resilience of these agri-industries;
- to explore what might be done in terms of local government planning in relation to the environmental and economic pressures facing the region which could help this vulnerable but essential peri-urban farming community.

Methodology

The case study area was the ridgeline of Hawkesbury–Mount Tomah, which abuts and bisects the GBMWHA.

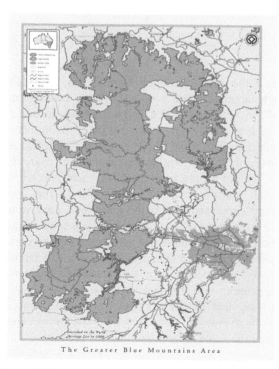

The Greater Blue Mountains Area

Figure 1 | The Sydney basin and GBMWH area

The area is essentially a peri-urban agricultural region dominated by orchards and some specialist organic and permaculture based farms. Land use is divided between economically active farms and 'rural lifestyle' large suburban blocks formed from subdivided farms.

Figure 2 depicts the overall project structure which was carried out in two stages. In stage 1, a literature review was undertaken that informed the subsequent stage 2, and which identified key economic, social and environmental issues operating at a range of temporal and spatial scales.

In stage 2, a series of 20 semi-structured interviews were conducted with a range of producers in the study area. These provided a basis for evaluating the issues faced by local producers, the majority of whom are orchardists. It identified two communities of practice in the area, one of which focused on local marketing and the other had a mainstream commercial marketing focus. Four farms were then selected for detailed case studies on the basis of their being representative of the range of farming systems within these two communities of practice. As one of the objectives of this study was to examine the regulatory, production and market factors associated with peri-urban agriculture in the region, it was decided to test a number of

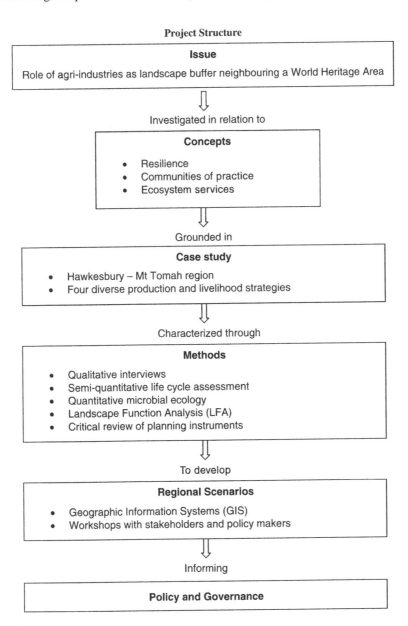

Figure 2 | Project structure

innovative management tools which might increase productivity, as well as economic and environmental sustainability.

Three tools were identified from a critical review of local literature, and the needs of farmers which emerged through the semi-structured interviews. Important selection criteria were their relevance to producers with different production and marketing strategies.

1. *Organic waste conversion (OWC).* A process was developed that converted orchard waste into a substrate for producing mushrooms and a disease-resistant mulch. Currently, when old orchard trees are removed the residue is burnt. Test conversions were undertaken on one of the case study properties with very positive results.
2. *Landscape function analysis (LFA)* was tested as a tool for comparing how different forms and stages of orchard production, semi-urban and urban land uses impacted on the three indices generated by LFA – slope stability, water infiltration and nutrient cycling. This application of LFA was used to shed some light on the ecosystem services generated by different forms of land use, such as orchard and urban development, and to link specific production systems to implications at a regional landscape scale.
3. *Geographic information systems (GIS).* A GIS application was explored that incorporated spatial and temporal information in relation to regional scale land use. Due to the space constraints of this paper the use of GIS in relation to OWC, LFA and climate change data is not included. It can however be found in the original report (Merson *et al.*, 2009).

Results of stage 1: preliminary analysis

Drivers of change affect agri-industries on a range of temporal and spatial scales (Table 1). At a societal and international scale, processes of land-use change and urbanization contribute to the fragmentation of landscapes, reducing their effectiveness in absorbing the impacts of economic and market changes. Policy responses to counteract this would recognize and enhance landscape functions and buffers, such as those provided by urban and peri-urban agriculture. At a regional scale, these change processes are expressed as differential real estate values, and changes in market access for local products. The values marginalized in the process include established agri-industrial livelihoods and ecosystem services generated by these land uses. Policy needs to identify and support agri-industries that contribute positively to ecosystem services. At a local scale, these combined drivers contribute to a breakdown in the viability of rural and regional townships, along with the character of the people and their sense of community. Integrative strategies already emerging include the diversification of production and marketing. Within households, financial viability is affected, reducing lifestyle and economic choices, with diversified sources of household income becoming the necessary strategy.

The preliminary analysis informed the range of methodologies for the study as shown in Table 2.

Review of background literature: land-use and regulation in the Sydney basin

Australia has not developed a deeply embedded tradition of retaining rural lands beyond its cities as

Table 1 | **Drivers of change in agro-ecosystems**

Scale	Dominant forces	Values marginalized	Potential integrative policies or strategies
Societal/ international	Land-use change; encroaching urbanization	Fragmentation of landscape and regional climatic processes	Role of agro-ecosystems as landscape buffers
Regional	Real estate prices; changes in market access	Agri-industrial livelihoods and ecosystem services generated	Regional role of agri-industries, particularly in relation to ecosystem services
Local	Economic viability of rural townships	Sense of community and character	Alternative production and marketing strategies; government–community partnerships
Household	Financial viability	Lifestyle choices and economic options	Diversified local livelihoods

Table 2 | **Potential supporting strategies at a range of temporal and spatial scales**

Key cluster	Potential strategies
Farm family/ business	• Whole farm planning • Enterprise facilitation
Spatial arrangements	• Demonstrations of best management practices • Catchment management/total water cycle management • Cumulative impact assessment • Ecosystem services • Geographic information systems
Communities of practice	• Information brokerage • Action research
Product	• Life cycle assessment • Waste minimization and recycling
Enabling policy	• Critical review • Community–government partnerships

valued economic, ecological and social resources. They have not been accorded the same status in legislation, planning or the collective community consciousness as lands within National Parks, World Heritage Areas or even urban open spaces and parklands within our cities. They have generally been regarded as 'lands in waiting' for some other higher or more pressing purpose such as industrial, urban or peri-urban development. As a consequence 'agriculture' undertaken on these lands has been historically regarded as a 'transient land use'. This is at the heart of the challenge faced by agriculture in and around the Sydney basin.

Over the past decades, a number of studies and workshops have discussed the challenges faced by agriculture in and around the Sydney Basin (e.g. Hawkesbury City Council, 1997, 2005; Kelleher *et al.*, 1998; Sinclair *et al.*, 2004; Mason and Docking, 2005; Knowd *et al.*, 2006). This section

introduces the key challenges identified in these studies, along with emerging adaptive strategies.

The need for a strategic approach to planning for agriculture as a critical component of the expansion of our cities has been clearly recognized for over a decade (Bakker *et al.*, 1999). Although the general attitude has been of traditional agriculture as a transient land use, there is now a growing call to better understand the multiple benefits of urban agriculture in the Sydney basin, the complex issues regarding the retention of agriculture, and the need for more creative and adaptive planning. In the Sydney region there has been an almost unstoppable trend towards the alienation of prime agricultural land as a result of urban encroachment and rural residential development. According to Kelleher *et al.* (1998, p. 4), this trend is adversely affecting the state's agricultural resource base, and 'agricultural land use studies by local government typically take an urban planning perspective, with an apparent tacit acceptance that rural residential subdivision will eventually occur'.

Kelleher *et al.* argue for the conservation of agriculture on the peri-urban fringe on the grounds that there is considerable evidence to support its importance economically as well as in terms of its protection of catchments and preservation of environmental and scenic amenity.

> Agriculture adjoining parkland, however, fills an important role in the Hawkesbury landscape by buffering parkland from the impact of urban development. It provides a transition zone in which the visual impact of urban development is reduced and it can provide important environmental services, such as water quality protection and air quality maintenance. Agricultural land also provides an ecological buffer and can act as a refuge and protective zone for wildlife (Kelleher *et al.*, 1998, p. 76).

Sydney's Metropolitan Strategy (DIPNR, 2005) states that greater recognition will be given to non-urban land so that it is not treated as land 'in waiting' for urban development. It is instructive to note, however, that the description of these lands as 'non-urban' reinforces the assertion by Kelleher *et al.* (1998) that agricultural land-use planning tends to be framed through an urban planning paradigm, which does not take into account the cultural and conservation imperatives associated with the GBMWHA.

Notwithstanding its reference to non-urban land, Sydney's Metropolitan Strategy does reflect the significant emergence of 'new recognitions' regarding the economic, ecological and social importance of agriculture in the Sydney Basin, and a greater institutional preparedness to respond to the complex challenges facing agriculture in more adaptive and sophisticated ways.

The 1998 Strategic Plan for Sustainable Agriculture in the Sydney Region (NSW Agriculture, 1998) has also played an important role in the development of this new recognition and response. Of equal significance is the emergence of local advocacy initiatives that reflect the agricultural community's recognition that a broader, more integrated community network approach is essential to promote agricultural products, influence policy and planning, and improve consumer awareness of the multiple values and benefits of agriculture. According to Mason and Docking (2005), the overarching goal is to provide an economic, social and environmentally sustainable agricultural industry that has wide community and sectoral support. Significantly, this integrated community network approach has the potential to be far more significant if it catalyses more informed community discourse around (1) the value of local agriculture in a carbon-constrained economy (including concepts such as 'food miles'), and (2) the strategic importance of local agriculture in terms of minimizing disruption to food supply in the event of crop failures in other areas (through drought, hail, frost and other climatic events).

Farming diversification, clustering and network development

Hawkesbury City Council initiated the 'Hawkesbury Agricultural Retention through Diversification and Clustering' (HARtDaC) project to address agricultural opportunities in the region which could assist in the retention of agriculture (Hawkesbury City Council, 2005). This project investigated options for farming diversification and clustering, and opportunities to enhance agricultural activity through farming networks. The HARtDaC project identified some key issues impacting upon agricultural retention. These included: the high comparative price for land with subdivision potential compared with land used for agriculture; reducing terms of trade associated with increasing efficiencies in food production and decreasing average lot size; the potential for escalating conflicts within the community, particularly with respect to noise, dust, water and odours; the role of changes in density of occupation and subdivision, and its influence on land use conflicts, rural amenity, regional tourism and natural resources; and long-term land degradation caused by inappropriate land management practices.

The HARtDaC project investigated strategies for agricultural retention that addressed socio-cultural, politico-administrative and environmental dimensions. In terms of the socio-cultural context, the key problem was interpreted to be the lack of awareness of the contribution of agriculture, together with intensive subdivision characteristic of some parts of Hawkesbury, which could lead to increased land-use conflicts and 'if reflected in the management of the region, may also result in inappropriate forms of governance' (Hawkesbury City Council, 2005, p. 153). Regarding the politico-administrative dimension, the HARtDaC study reflected a very complex regulatory structure resulting from the multi-tiered political system. Local stakeholders highlighted three key concerns: 'unclear regulatory structure; regulations and processes not informed by agriculture; and the need for more support for agricultural innovation during the planning process' (Hawkesbury City Council, 2005, p. 156). For the economic decision context, the primary issue was that while agriculture contributes considerably to the regional economy, peri-urban agriculture was characterized by reducing viability associated with increases in land values, ongoing reduction of farm sizes, and therefore an increased reliance by most farm families on off-farm income. This conflict in economic values between agricultural production and urban development has been addressed in the United States by Heimlich and Anderson (2001).

Land use planning

Sinclair (2001a) divided rural residential development into two parts: *the rural urban fringe*, or development that is within the servicing catchments and located close to the urban centre, and *rural living*, or residential use of land within a rural environment. Both types use rural land for residential, as opposed to agricultural purposes, and can be distinguished from urban housing by the larger lot sizes and distance between dwellings. Rural residential development is increasingly common on the fringe of metropolitan areas and Hawkesbury City Council is typical of many local councils in that it is required to find ways to deal with this very complex local planning issue. Sinclair (2001b) argued that rural residential development can have both positive and negative impacts on an area. Positive impacts include lifestyle choice, provision of land for businesses needing space for storage, and potential contribution to the land economy. These are outweighed by such negative impacts, in Sinclair's view, as: the increased financial costs of a scattered settlement pattern; community costs relating to provision of services and facilities located at a distance from town centres; and environmental costs connected to the initial development (for example, clearing of native vegetation, soil erosion and land degradation). In addition, problems associated with the ongoing use of the land include the impacts of onsite effluent disposal, soil and water management, weed invasion and domestic pets (Sinclair, 2001b).

Given the above experience of issues resulting from intensive agriculture meeting rural residential living head on, it is clear that land-use planning, particularly with reference to lot sizes, subdivision and zoning objectives, is of paramount importance in maintaining agricultural land on the urban fringe. There is an abundance of international, national, state and local land use regulations that have bearing on the land-use planning process, especially in areas that presently act as a buffer to the GBMHWA. Local government authorities are required to juggle the competing interests of those seeking a rural residential lifestyle and those attempting to maintain the agricultural productivity of the peri-urban fringes.

The local planning context in Sydney is complicated by two particular dimensions, these being local emerging land-use planning within the local political context, and state-wide pressures to standardize local governmental environmental planning instruments. A number of councils are developing useful steps toward responding to their particular local situation, but this has occurred in the context of local polarities in perspectives towards development.

Transforming urban agriculture

Development of a new urban agriculture results from a significant transformation in action and thinking. New product and marketing strategies develop, along with recognition of the ecosystem services provided by agricultural landscapes. Concepts of agro-ecosystem resilience are emerging as useful tools to inform and guide transformational change in agricultural enterprises, industries and landscapes. These transforming processes have set the context for the development of this study and informed its methodology and implementation. Knowd *et al.* (2006) identified a number of key transforming themes emerging worldwide, including local food, direct marketing, innovation and adaptability of urban agriculture, the urban agriculture/public health relationship, and agricultural land preservation. The diversified nature of differing forms of urban agriculture generates broad-ranging economic, environmental and social values. However, the classic problem is that many of these intrinsic social and environmental values are not adequately reflected and

accounted for in formal institutional, market and decision-making arrangements.

This new urban agriculture is described by Butler and Maronek (2002) as leading to a range of other benefits and services including recreation and leisure, economic vitality and business entrepreneurship, individual and community health and well-being, landscape beautification, and environmental restoration and remediation.

Intensification of agriculture in the Sydney Basin has not been the result of strategic intervention by government or industry groups, but rather the adaptive and opportunistic responses to market requirements and the changing socio-economic situation (Mason and Docking, 2005). Threats to some industries relate more to industry issues (such as de-regulation of the dairy industry) or to external factors beyond regional control than to subdivision or urban encroachment. However, urban encroachment is the single greatest threat to the most economically important industries in the Hawkesbury to date (Kelleher *et al.*, 1998).

The principle agricultural industries identified by Kelleher *et al.* (1998) in the Hawkesbury Local Government Area (LGA) were mushrooms, turf, fruit, market gardening and dairy. Their study suggested that the industries of greatest economic importance were also those vulnerable to the impacts of urban expansion. Worldwide, small area farming and agri-industries continue to develop new means of diversification (van Veenhuizen, 2006). These include niche markets and new agri-industrial configurations, and clusters of local produce sold directly to consumers are becoming more common. This provides a greater proportion of the consumer dollars to the producer, along with the social benefits of increased communication and understanding across the urban–rural divide. With consumer tastes and demands also driving larger agri-industries, these direct marketing options provide a means for greater expression of consumer preferences for production strategies that are environmentally and socially responsible (van Veenhuizen and Danso, 2007). Alternative strategies mentioned in the HARtDaC study include: developing new skills to incorporate tourism, recreation and related value-adding to produce; alternative systems such as permaculture and organic produce; and cooperative marketing and supply chain management.

A local example is that of Hawkesbury Harvest (HH), which seeks to promote better community access to locally grown food, enabling the opportunity for the diversification of income. Mason and Docking (2005) describe the Hawkesbury Harvest model as encompassing industry clustering, industry development, small business development, income generation, community gardens, controlled environment intensive horticulture, matching local climate to crops and markets, farmers markets, agri-tourism, research and education, and training through extension. The overarching goal of HH is to provide an economic, social and environmentally sustainable agriculture industry that has wide community sectoral support.

Agricultural ecosystem services

Ecosystem services are described by Daily (1997) as the conditions and processes through which natural ecosystems, and the species that make them up, sustain and fulfil human life. On the basis of this definition, ecosystem services are a way of thinking about the fundamental ecological processes and capacities that enable our economies and societies to operate. This study sought to investigate the important role peri-urban agriculture plays in maintaining ecosystem services as landscape buffers around areas of significant natural and cultural heritage, such as the GBMWHA. A critical issue for this study was that while the goods generated by agri-industries are accounted for and economically valued, the broader environmental services that are generated tend not be to accounted for and therefore are not explicitly incorporated in policy and planning processes.

Some ecosystem services can be considered as 'umbrella services' supporting a nested hierarchy of other services which are contingent upon them. One example would be those functions and processes under the rubric of soil health. These reflect the ecosystem functioning of soils and support many of the buffering mechanisms and transformations. In the investigation of ecosystem services in the Goulburn Broken Catchment by the Australian Commonwealth Scientific and Research Organisation (CSIRO), soil management was identified as perhaps the single most significant on-farm ecosystem service issue in the catchment (Binning *et al.*, 2001).

While studies of ecosystem services have been undertaken at a range of scales, the key interest of this study was the nature of these services generated on-site within agri-industries that contribute to regional and landscape functions. It is recognized that there is an interdependent relationship between local and broader scales. The general ecosystem services described by Cork *et al.* (2002) include: pollination; life fulfilment; regulation of climate; pest control; provision of genetic resources; maintenance of habitat; provision of shade and shelter; maintenance of soil health; maintenance of healthy waterways; water filtration and erosion control; regulation of rivers and groundwater; and waste absorption and breakdown.

A study of agricultural landscapes by Swift *et al.* (2004) was found to be particularly useful in providing a framework for understanding the more local aspects of ecosystem services generated by well-managed agro-ecologies, with particular focus on soil and microbial roles. The complexity of interactions between tolerances in ecosystems and the driving processes of markets and other institutions for planning, management and governance is becoming well recognized (e.g. Cork *et al.*, 2002), but it remains difficult to deal with methodologically. Previous studies such as those undertaken by the CSIRO (e.g. Binning *et al.*, 2001; Cork *et al.*, 2002; Abel *et al.*, 2003) recognized the importance of approaches that combine case studies and participation with a suite of varied supporting analytical methodologies. Due to the nature of this complexity, the methodological approach taken in this study was adaptive, beginning with broad qualitative means of investigating the dimensions of the situation, and refining particular methodological means to support key areas of potential advocacy that emerged.

Results of stage 2: participation of primary producers

The second stage of this study was to invite the participation of farmers who represented the diversity of production and marketing strategies found in the area (Table 3). Based upon the initial series of semi-structured interviews, four different primary producers

Table 3 | Horticultural producers involved in the project

Producer	Enterprise description
A. Shield Orchard	Apple orchard, pick your own and roadside stall sales
B. Saliba Fruits	High productivity commercial apple orchard supplying Sydney markets and contracted to major retailer
C. Inniskillen Orchard	Small-scale mixed orchard, retail outlet for regional produce, café business
D. Chorley's Farm	Permaculture farming and berry production supplying specialist and growers' markets

were invited to be involved as representative demonstrations and case studies. A letter was sent out, inviting ongoing participation with the intention of holding further focused discussions to address:

1. Gaining a detailed understanding of the strategies applied by the family business, including:

   - product life cycles and supply chain strategies;
   - interactions in the local landscape.

2. Opportunities to enhance the recognition of good practice, and to address issues such as:

   - value-adding from waste streams and second grade products;
   - recognition of ecosystem services.

An initial characterization of each production and marketing system was undertaken through adapting a life-cycle assessment to the operations undertaken for each step of their production and marketing strategies. As outlined in Figure 3, inputs and outputs were investigated in relation to the following life cycle stages relevant to orchard production:

- land preparation and establishment, including infrastructure;
- maintenance and orchard productivity;
- product harvest and alternate product sources;
- processing and product differentiation;
- marketing, packaging and transport.

The four farmers who agreed to participate in the project represented different types of horticultural production systems. They agreed to participate in a review of the waste streams generated from their farms, and to be involved in evaluating the outcomes of this project. An informal assessment was undertaken to identify types of waste, waste volumes, waste production periods and current methods of waste management. The waste streams identified represent those associated with several different types of horticultural production and are of relevance to both traditional horticulturalists and the exponents of permaculture methods. All the horticulturalists expressed an interest in waste management issues, but from different perspectives, and this was considered when attempting to match waste treatment to the requirements of growers. As a consequence, it was decided to apply three integrated tools, each operating at different spatial and temporal frames and each of which would provide new products and environmental management processes.

Organic waste conversion

The aim of the organic waste conversion (OWC) element of the study was to identify suitable microbial bioremediation agents for several horticultural waste

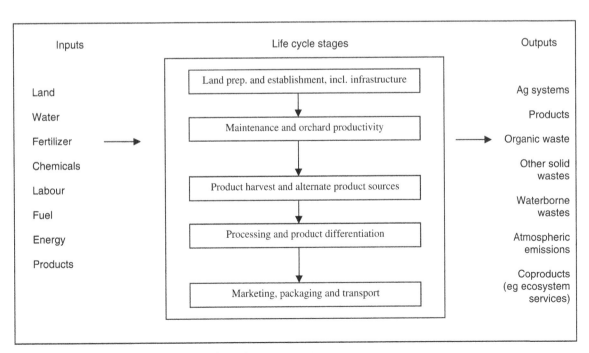

Figure 3 | General structure used to inform life cycle assessments

streams in the Hawkesbury/Bilpin region and to investigate the conversion of these wastes into valuable economic and environmental resources. Vast quantities of organic wastes, particularly those containing lingocellulosic materials, are generated through primary and secondary production systems in the agricultural, forest and food-processing industries. Currently, large proportions of these wastes are either burnt or go straight to landfill with resultant economic and ecological implications. For small-scale horticultural businesses, the generation and disposal of waste streams may impact upon profits and have serious ecological consequences for the local environment. However, appropriate bioremediation can convert these wastes into valuable economic and environmental resources. A detailed assessment of waste types, volumes, production periods and current methods of waste management were undertaken in a pilot study carried out at one of the orchards. On the basis of this, apple tree waste was selected as a trial substrate for the growth of fungi, since fruit tree removal and destruction was identified as the likely source of most on-farm waste by all producers.

The study was designed to explore the potential for using microbial processes to break down waste material and at the same time produce useful commercial end products. In this particular case, organic waste was chipped and converted into a substrate for the growing of oyster and shiitake mushrooms which have high market value. The spent substrate was then used as a mulch for the orchard trees. This mulching improves the soil condition and due to the bacterial process involved in the mushroom production it has been found to increase resistance to nematodes which attack the roots of orchard trees. The approach has now been adopted by Bill Shield on farm A, and collaboration with the research specialists involved is continuing. As the technique becomes more widely recognized it is anticipated that it will be taken up by other orchardists as a means of addressing organic waste as burning becomes prohibited, and also for its bacterially useful mulch and the commercial value of specialty mushrooms involved.

Landscape function analysis

Landscape function analysis (LFA) is a method for assessing the ability of a slope to retain and utilize its vital resources of water, soil and litter. It involves site analysis and the division of a down-slope transect into zones according to whether they retain (a patch) or lose resources (an interpatch). A soil surface assessment of each zone is then undertaken (see Figure 4). The outputs are numerical indices of key landscape functions for each zone and for the whole slope: stability, water infiltration and nutrient cycling. A low stability value indicates that the slope shows evidence of soil loss, a low water infiltration value showing that water is not soaking in and being used by the vegetation but is running off the slope, and a low nutrient cycling value indicates that there is evidence of lack of perennial plant growth, litter accumulation and decomposition. High values indicate that soil is well protected and retained, that most water is soaking in, and that nutrients in the slope are being actively and extensively cycled.

LFA has been developed over three decades by a number of CSIRO scientists predominantly on rangelands and in mine reclamation work (Tongway and Hindley, 2005, p. 14). It is a systematic approach to using 'visually assessed indicators of soil surface processes' to demonstrate whether a slope is leaking or retaining water, soil, litter and nutrients. Done in time sequence it can track the impact of land-use change and soil remediation strategies. If replicated across a particular land type, LFA can provide evidence of the dynamic range of the land types and can identify the critical point above which the slope is on a trajectory to self-regeneration, and below which effort is needed to prevent degradation.

In this study we used LFA on each of the four farms to compare the leakiness of different stages in the orchard cycle with one another, and with other land-uses on the same property (Table 4). In addition, LFA was undertaken on a suburban house block to generate data that could be used to assess the likely impact of urbanization on areas that are presently under orchard and other agricultural land uses (Figure 5).

This work provided an initial determination of the utility of LFA as a method for comparing the functional attributes of different land uses. Comparisons can confidently be made between zones on the one site, but values between sites can only be compared cautiously. Where differences in values between sites are large, conclusions need to be made in the context of key features represented at each of the sites.

LFA discriminated quite finely between orchards with different ages of trees, under tree mulch and between row vegetation. It also provided comparative data that could be used to contrast the functionality of, for example, grassy slopes, lawns, bushland and orchards on the same or similar hill slopes. LFA did not appear to discriminate between land uses that were already functioning at a quite high level based on landscape function indicator values for stability, infiltration

LFA transect running down a slope previously under an orchard

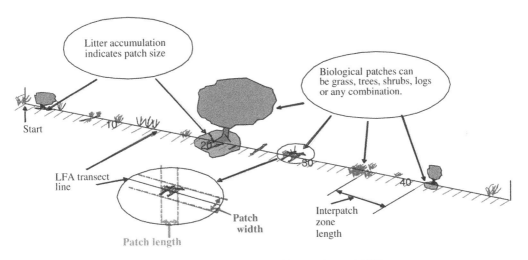

Figure 4 | Landscape organization (showing key features involved in carrying out LFA)

and nutrient cycling (e.g. for a multi-species permaculture garden and a mature, well-mulched orchard with a thick stand of perennial grass in the inter-row). The application of LFA to a suburban house, garden and street would appear to be quite informative, especially if it allows urban areas to be included in across-landscape comparisons. As LFA generates standard errors, a more comprehensive study would generate sufficient data to determine whether differences in LFA values between land uses are statistically significant. There were insufficient resources to do this in this study.

Integration of the tools

The value of LFA as a farm-level tool and as a means of assessing and monitoring changes in land use at a regional level was recognized by the farmers involved in the field trials at Fields Orchard (A) in 2008. LFA has usually been applied to environmental assessments

Table 4 | **Summary of landscape function analyses conducted on case study sites**

Producer	LFA comparisons	Tentative conclusions
A	Mature and former orchards were compared; both were divided into under-row and between-row patches	Removing mature trees reduces landscape function and increases leakiness
B	Two orchard types and components of a typical parkland were compared on the same slope	A bush patch functioned better than orchards and other components of parkland, except for dense perennial grass inter-rows; young orchard functioned lowest
C	Orchard, former orchard and backyard were compared on same slope	Orchards with mulch under the trees had similar function to a mown back yard; unmulched orchards were leakier
D	Different components of a permaculture garden were compared	All elements of the garden functioned highly apart from where ground under trees was heavily shaded
Suburban block	Different components of an established garden were analysed, including lawns, garden beds containing annuals and perennials, paths, driveways and the house; these values were extrapolated to a typical house block	Areas dominated by perennials functioned highest; lawns, driveways and houses dramatically reduce function

in rangelands and for land being reclaimed after mining. However, coupled with the application of soil microbial systems at one end of the spectrum and GIS at the other, a suite of interlocking tools for monitoring environmental change in each study area could be developed which would act at a complementary range of temporal and spatial scales.

LFA also has potential application at a regional level through its use in conjunction with information generated by GIS. GIS tools with the capacity to operate at both the farm and regional level also allow for the monitoring of the results from changes in farm and land management practices. These might include such environmental goals as maximum water retention and minimal erosion and leakiness. In this respect GIS analysis provides the capacity to model and evaluate these environmental and ecological processes of landscapes at a regional level. GIS can be used to provide a

regional framework for LFA through providing a catchment-wide model for biophysical processes. It is useful for researchers/practitioners to be able to relate LFA to these regional scale indices. Within an adaptive management approach there is also potential to develop thresholds of potential concern (TPCs) for a catchment, and to monitor the contribution of each private landholding to the catchment's health, e.g. remaining below specific thresholds based on LFA measurements.

A package of tools using existing Arcview and Arcgis extensions was also developed to support this project, along with relevant GIS data. This system used spatial data such as patch and hydrological analysis, along with erosion and depositional modelling. Several models were produced to provide an indication of the application of GIS to LFA in this project. It is anticipated that the GIS modelling tool developed for the region can also be used to assess climate change

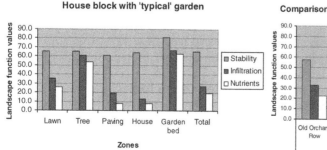
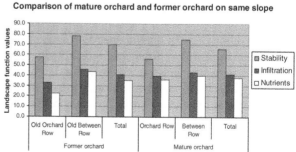

Figure 5 | A typical suburban house block compared with an orchard

impacts. As data reliability improves, it can be used as a decision support system to assist farmers in assessing the long-term viability of crop types in relation to predicted changes in temperature and rainfall. However, these applications were not able to be fully tested within the scope of the project and are now part of an ongoing research project.

Conclusions

In the process of identifying agri-industries existing among the diverse landholdings, and documenting their economic, social and environmental impacts, this project has confirmed that despite the tacit support from local government, urban agriculture in this region is under considerable pressure. The establishment of the Hawkesbury Harvest has been significant for providing support for the marketing and branding of regional products, but more initiatives are needed. Despite the recommendations of the Hawkesbury City Council's HARtDaC report (2005), local government planning remains confused and contradictory in terms of the support for and retention of urban agriculture in the region. Nonetheless the very diverse modes of production as exemplified in the four different farming operations reviewed suggest that there is still potential resilience in the production systems. However, farmers and regional land management agencies need more innovative tools and strategies to address the challenges of increasing productivity, the pressure of urban development, as well as changing climatic conditions and the environmental need for low emission production systems. In this respect the toolsets developed as part of this project provide a modest starting point.

The project recognized at the outset the need to address the complex social, economic, technical and political variables that govern the region's agri-industries. This led to the recommendation that a more adaptive strategy based on the used of the tools outlined above would have a dual benefit. An integrated biosystems approach to waste using the microbial conversion of old apple trees could support more sustainable farming practices and at the same time increase farm income through mushroom production. Monitoring across farms using LFA would help address the 'leakiness' of systems (e.g. erosion, loss of nutrients and inefficient water management), and could be used to predict impacts of land-use change and test strategies for remediation. It would also allow for a more objective assessment of urban and rural land-use practices. GIS tools could in turn provide an aggregation effect of these remedial strategies at a catchment and regional level.

However, these strategies also need to be supported by local and state government through more targeted land-use policies and regulation, especially if this peri-urban agriculture sector is to continue to provide its traditional economic and ecosystem services to the regional community. There is also a good argument for some form of financial subsidy to be provided by the state to support the critical role that urban agriculture provides as a buffer zone between the suburban expansion of Sydney and the fire-dependent bushland areas of the GBMWHA. This is now a critical issue as climate change evidence and models demonstrate an increased risk of more intense and frequent bushfires, with the potential of their moving unhindered from the World Heritage area into urban areas and vice versa, if the present buffer zone functions of peri-urban agriculture is replaced by urban development. The issues faced by farmers in the Hawkesbury Nepean and Blue Mountains regions of Australia are comparable to those confronted by many urban agricultural communities. It is hoped that the approach taken in this project will be of use both within Australia and in other countries.

Acknowledgements

The authors acknowledge the Rural Industries Research & Development Council (RIRDC) for funding this project, and the farming families who participated in the study.

References

Abel, N., Cork, S., Goddard, R., Langridge, J., Langston, A., Plant, R., Proctor, W., Ryan, P., Shelton, D., Walker, B., Yialeloglou, M., 2003, *Natural Values: Exploring Options for Enhancing Ecosystem Services in the Goulburn Broken Catchment*, CSIRO Sustainable Ecosystems, Canberra.

Bakker, N., Dubbeling, M., Guendel, S., Sabel-Koschella, U., de Zeeuw, H. (eds), 1999, *Growing Cities, Growing Food: Urban Agriculture on the Policy Agenda: A Reader on Urban Agriculture*, Resource Centre on Urban Agriculture and Forestry, Leusden, The Netherlands.

Binning, C., Cork, S. J., Parry, R., Shelton, P., 2001, *Natural Assets: An Inventory of Ecosystem Goods and Services in the Goulburn Broken Catchment*, CSIRO Sustainable Ecosystems, Canberra.

Bradstock, R., Williams, J., Gill, M., 2002, *Flammable Australia: The Fire Regimes and Biodiversity of a Continent*, Cambridge University Press, Port Melbourne, Australia.

Butler, R., Maronek, D. M. (eds), 2002, *Urban and Agricultural Communities: Opportunities for Common Ground*, Council for Agricultural Science and Technology, Ames, IA.

Cork, S. J., Proctor, W., Shelton, D., Abel, N., Binning, C., 2002, 'The ecosystem services project: exploring the importance of ecosystems to people', *Ecological Management & Restoration* 3 (2), 143–146.

Daily, G. (ed.), 1997, *Nature's Services: Societal Dependence on Natural Ecosystems*, Island Press, Washington, DC.

Dang, H. T., Malcolm, P., 2006, *Improved Economic Sustainability of Vietnamese Vegetable Growers in the Sydney Basin*, Final Report. Project No. DAN 211A, RIRDC, Canberra.

DIPNR, 2005, *City of Cities: A Plan for Sydney's Future*, The Metropolitan Strategy, NSW Department of Primary Industries and Natural Resources [available at www.metrostrategy.nsw. gov.au/dev/uploads/paper/introduction/index.html].

Hawkesbury City Council, 1997, *Healthy Cities Survey*, Hawkesbury City Council, Windsor.

Hawkesbury City Council, 2005, *Growing Hawkesbury's Future – the Hawkesbury Agricultural Retention through Diversification and Clustering Project (HARtDaC)*, Agriculture Victoria Services, Windsor [available at www.hawkesbury.nsw.gov.au/ environment/1232/21007.html].

Heimlich, R., Anderson, W., 2001, 'Development at and beyond the urban fringe: impacts on agriculture', *Agricultural Outlook* August. Economic Research Service/USDA, Washington, DC.

Kelleher, F. M., Chant, J. J., Johnson, N. L., 1998, *Impact of Rural Subdivision on Agriculture*, A Report for the Rural and Industrial Development Corporation, Farming Systems Research Centre, University of Western Sydney, Hawkesbury.

Knowd, I., Mason, D., Docking, A., 2006, *Urban Agriculture: the New Frontier*, Paper presented to the Planning For Food Seminar, Vancouver, 21 June 2006.

Mason, D., Docking, A., 2005, *Agriculture in Urbanising Landscapes: A Creative Planning Opportunity*, Paper presented to Agri-Food '05, 5–8 July 2005, Yeppoon.

Merson, J., 2006, *An Assessment of the Potential Impact of Climatic Change on the Greater Blue Mountains World Heritage Area*, UNESCO, World Heritage Centre, Paris.

Merson, J., Attwater, R., Booth, S., Mulley, R., Ampt, P., Wildman, H., Nugent, M., Hooper, S., Campbell, M., Chapple, R., 2009, *Urban Expansion and Sensitive Environments: Assessing the Role of Agri-industries as Landscape Buffers to the Neighbouring Greater Blue Mountains World Heritage Area*, RIRDC, Canberra.

NSW Agriculture, 1998, *Strategic Plan for Sustainable Agriculture – Sydney Region*, NSW Agriculture, Orange.

Sinclair, I., 2001a, 'A view from the edge. Issues in rural and metropolitan fringe planning: rural residential development', *New Planner* 47.

Sinclair, I., 2001b, 'A view from the edge. Issues in rural and metropolitan fringe planning: rural residential development impact', *New Planner* 48.

Sinclair, I., Docking, A., Jarecki, S., Parker, F., Saville, L., 2004, *From the Outside Looking In: The Future of Sydney's Rural Land*, University of Western Sydney, Hawkesbury, Australia.

Swift, M. J., Izac, A. M. N., Van Noordwijk, M., 2004, 'Biodiversity and ecosystem services in agricultural landscapes – are we asking the right questions?', *Agriculture, Ecosystems and Environment* 104 (1), 113–134.

Tongway, D. J., Hindley, N. L., 2005, *Landscape Function Analysis: Procedures for Monitoring and Assessing Landscapes*, CSIRO Sustainable Ecosystems, Canberra.

Van Veenhuizen, R., 2006, *Cities Farming for the Future – Urban Agriculture for Green and Productive Cities*, RUAF Foundation, IDRC and IIRR.

Van Veenhuizen, R., Danso, G., 2007, 'Profitability and sustainability of urban and peri-urban agriculture, Agricultural', Management, Marketing and Finance Occasional Paper 19, FAO, Rome.

Agronomic considerations for urban agriculture in southern cities

Nikita Eriksen-Hamel[1]* and George Danso[2]

[1] Canadian International Development Agency, Ottawa, Canada
[2] Department of Rural Economy, University of Alberta, Edmonton, Canada
The views expressed in this publication are those of the authors and do not reflect the views of the affiliated institutions.

Urban and peri-urban agriculture (UA) provide a significant contribution to the total food requirements of cities, especially in southern cities of the developing world. Increasing food production in UA is therefore a necessity for increasing the food security of the urban poor. Urban environments are inherently different from rural environments and these differences in environmental conditions are expected to impact differently on crop growth. This review describes agronomic issues that are unique to UA and identifies possible interventions to address them. The constraints that can significantly differ include temperature, air quality, solar radiation and climate. The growth-limiting and growth-reducing factors that affect actual production in UA include water availability, nutrient supply, soil degradation, pests and soil pollution. The interventions addressing these constraints require action at both field level, and municipal or regional levels. The food security of the urban poor will therefore require coordinated efforts and cooperation between the farmers who produce food and the planners and policy makers who manage the supporting systems such as markets, inputs and land registration.

Keywords: crop production, production constraints, theoretical and potential production, urban agriculture

Introduction

Food production in urban and peri-urban areas provides significant contributions to the total food requirements of cities. This food is channelled to urban families either through subsistence production or through commercial urban food markets. The commercial scale of food production largely depends on municipal zoning by-laws, environmental regulations, land, water and labour costs, and the availability of non-agricultural jobs (Cabannes and Dubbeling, 2001; Mougeot, 2006; Nsangu, 2008). For these reasons urban and peri-urban agriculture (UA) within wealthier northern cities is mostly focused on non-commercial agriculture practised on family-scale plots and for the purposes of improving the quality of diet and life (Bhatt, 2005). Some commercially oriented farms do exist within urban areas of northern cities but these exceptions are largely due to environmental reasons (greenbelts), special permits, or long-standing land rights (Bhatt, 2005) and can involve alternative agriculture or mixed-business enterprises.

In comparison, UA in poorer cities of the south involves a greater mix of commercial farms and family-scale production. Urban food markets are the main destination of food produced on commercial farms, while some of the food produced by families is often sold in neighbourhood markets to supplement family income (Maxwell, 1994). The contribution that UA makes to the food self-reliance and food security of a city varies considerably, but it is clear that in southern cities this contribution is substantial. Numerous studies of large cities in Asia, including Kathmandu, Singapore, Shanghai and Karachi, show that between 30 and 85 per cent of the cities' vegetable requirements are met by UA (Yeung, 1985; Wade, 1987). Similar studies for major cities in Africa, including Accra, Kumasi, Nairobi, Dar es Salaam and Dakar show up to 80 per cent of vegetable requirements are met by UA (Foeken and Mwangi, 2000; Jacobi *et al.*, 2000; Mbaye and Moustier, 2000; van Veenhuizen and Danso, 2007). Estimates for Kampala, Nairobi and Dar es Salaam show some 40 per cent of staple crops are produced within city limits (Jamal, 1988; Maxwell and Zziwa, 1992; Sawio, 1993).

*Corresponding author. Email: nikita.eriksenhamel@gmail.com
INTERNATIONAL JOURNAL OF AGRICULTURAL SUSTAINABILITY 8 (1&2) 2010
PAGES 86–93, doi:10.3763/ijas.2009.0452 © 2010 Earthscan. ISSN: 1473-5903 (print), 1747-762X (online). www.earthscan.co.uk/journals/ijas

The growing importance of food production in urban areas requires increased attention by all stakeholders to identify the constraints which impede the development and growth of UA. To this end, researchers have identified some of the major economic, land-tenure, social, waste and health constraints (Bryld, 2003; Afrane et al., 2004; Drechsel et al., 2004; Mougeot, 2006). A number of case-based studies have investigated the economic production chain (food to market) of UA as well as the social, economic and health benefits of urban food production (Veenhuizen and Danso, 2007). These provide justification for the benefits of UA and outline some of the major social, economic and institutional constraints to the development and practice of UA. Although the complexities and extent of UA are becoming better understood, there remains a gap in the literature regarding the agronomy of urban agricultural systems.

The agronomy of crop cultivation in urban areas should be similar to that of crops grown in rural areas. The specific objectives of this review are (1) to identify agronomic constraints related to potential and actual crop production that may be inherently unique to UA and different from rural agriculture, and (2) to identify possible interventions and solutions that may help urban farmers adapt to these constraints. We hope that identifying possible interventions for each constraint may stimulate discussions and encourage further research into these farming activities.

Constraints affecting urban crop production

Potential crop production is defined as the maximum output that a plant can achieve at an optimum supply of all inputs and in the absence of any growth-limiting conditions (van Ittersum and Rabbinge, 1997). The environmental conditions of a given location determine these growth factors. Since urban areas have environmental conditions that are distinctly different from rural areas we can expect differences in potential primary production of crops between urban and rural areas. These differences in environmental conditions include temperature, air and soil quality, solar radiation and climate or weather patterns, and are discussed below.

Solar radiation
Irradiance at certain sites around the world has decreased from the 1960s to 1990s (Stanhill and Cohen, 2001). These linear decreases have been significantly greater in urban areas than rural areas regardless of latitude (Stanhill and Cohen, 2001). Alpert and Kishcha (2008) later found that solar radiation decreased significantly in areas with high population density (>100 persons km^{-2}) and that this effect was more pronounced at low latitudes ($40°S–40°N$) where many developing country cities are located. The cause of solar dimming is explained through the increased reflectance of radiation away from the ground, due largely to increases in air pollutants and aerosols over urban areas. Heavily polluted urban areas receive 8 per cent less solar radiation than rural areas (Alpert and Kishcha, 2008). However, the impacts of solar dimming on the potential production of crops in urban areas are difficult to predict. UA is also exposed to relatively high short-wave radiation, reflected from buildings and paved surfaces, which is likely to create heat load and deplete soil moisture relative to what would be expected when only direct incoming irradiance is measured. Shading experiments on lettuce showed a proportional decrease in crop growth with a decrease in radiation (Sanchez et al., 1989), but the response of other vegetables or fruit crops is less clear (Mourão and Hadley, 1998; Cohen et al., 2005). The effect of decreased solar radiation on plant productivity is likely to be affected by the radiative heat load on the plants (Wang et al., 1994). In wet humid climates with low radiative heat load a decrease in plant productivity is likely if solar radiation is decreased.

An obvious adaptation to lower direct solar radiation within cities is the selection of species, shade-adapted ecotypes, or sunny locations. Shade-adapted plants may reach maximum photosynthesis at a lower saturating irradiance than plants of the same cultivar growing in full sun (e.g. 300 compared with 600uE for coffee: Kumar and Tieszen, 1980), and it has long been recognized that C3 plants, which include most vegetables and fruit trees, are more adapted to low irradiance than C4 species. Fortunately, C3 species and fruit trees are also more responsive to elevated carbon dioxide concentrations than C4 and herbaceous food crops (Ainsworth and Long, 2005). Thus, plant breeding for UA appears to have distinct and non-contradictory opportunities. There are also possibilities that plant growth in low light can be enhanced with targeted nutrition, such as additional calcium (Liang et al., 2009).

Air pollution
Rapid economic growth, industrialization and urbanization in southern cities has led to increases in air pollutants such as ozone (O_3), nitrous oxides (NO_x), sulphur dioxide (SO_2) and suspended particulate matter (SPM).

Carbon dioxide concentrations and air temperatures are commonly higher in cities, and wind movement lower. In urban areas, poor or uncontrolled burning of crop residues or municipal solid wastes, and the use of diesel fuel contributes to increased suspended particulate matter. Other pollutants originate from motor vehicles and industrial sources, both of which are significantly greater in urban areas than rural areas.

At high concentrations these gases cause significant damage to crops. Emberson *et al.* (2001) document instances of damaged crops and decreased yields in various cities. The impact of air pollution can be especially significant for fruits and tree crops. In China, high levels of air pollution cause damage to fruit trees by delaying sprouting, shortening the flowering period, accelerating senescence, reducing CO_2 assimilation, reducing fruit numbers, and premature dropping of fruit (Zheng and Chen, 1991; Boa and Zhu, 1997). High ozone concentrations cause chlorotic spotting and premature senescence in trees, vegetable crops and cereals (Rich, 1964; Krupa *et al.*, 2001). However, apart from ozone, it can be difficult to identify the causal link between a specific gas and damage to crops. Suspended particulate matter, for example, is expected to be most damaging not only in terms of decreased yields but also by reducing the quality of crops, especially fruits and leaf vegetables. Atmospheric deposition is the dominant pathway for lead contamination of leafy vegetables in Uganda arising from high traffic density (Nabulo *et al.*, 2006). The concentration of lead in the vegetables falls with growing distance from roads, and farmers are recommended to cultivate leafy vegetables at least 30–100 m from roads (Ward *et al.*, 1975; Rodriguez-Flores and Rodriguez-Castellon, 1982; Nabulo *et al.*, 2006).

The significant impacts that air pollution has on the respiration and photosynthesis processes of crops in UA reinforce the importance of addressing this agronomic constraint. However, few crop-wide mitigation options are available for farmers other than planting crops away from areas of high deposition. Interventions that improve the gas exchange and net photosynthesis of plant leaves such dusting or washing to remove particulate matter may be possible. Similarly, fine mesh netting over high-value crops could reduce large particulate matter and debris settling on crops, but there needs to be an economic benefit through improved food quality and price.

Urban farmers could also benefit from plant breeding if new cultivars were developed for the unique conditions of urban environments, particularly tolerance to shade and heavy metals. The development of specific 'urban' cultivars will depend on certain conditions. Seed companies must be convinced that urban farmers offer a distinct market for their products. The promotion of participatory plant breeding among urban farmers might also be an option but would require commitment for funding, training of urban extension officers with rural knowledge, and training of farmers and farmer groups.

Soil degradation

The risk of soil degradation in UA is dependent on the types of crops and location of cultivation. The use of urban land to address essential needs such as housing (residential), jobs (industrial and commercial), and social services (institutional) will almost always take precedence over UA unless strong land-use and zoning laws are in place. The informal urban planning of southern cities favours commercial, industrial and residential development on flat lands, and thus often forces UA to marginal lands on steep slopes, valley bottoms, or in areas adjacent to polluting industries or roads.

Enforced zoning for UA does occur in some northern cities, such as in the greenbelts of Ottawa, Toronto and Vienna, and designated municipal gardens in Montreal and London (Taylor *et al.*, 1995; Garnett, 1996; Vogl *et al.*, 2004). Although this type of legislation is not common in southern cities, it is gradually coming to the attention of policy makers. In Mexico City, land set aside for UA in nature reserves bordering the city was either legally expropriated by land developers or illegally invaded for low-income housing, forcing UA into more marginal lands further away from the city core (Torres-Lima *et al.*, 2000). In 2010, UA in Villa Maria del Triunfo in Lima, Peru will be recognized by all stakeholders as a dynamic and integrated activity that contributes to sustainable development (RUAF, 2007). Similar programmes of integrating UA into city development plans can be found in the RUAF-CFF programme (www.ruaf.org). However, where integration does not occur, social and economic pressures force agriculture into lands that are inherently more at risk of landslides or floods (Douglas *et al.*, 2008).

Adaptations to soil degradation risks will involve physical effort, financial investment and increased farm management. Infrastructure interventions, such as terraces and check-dams, can also be important parts of successful soil protection plans. However, the insecurity of access and tenure of land, a common reality for urban farmers, may limit the use of these interventions. Thus urban farmers may need to concentrate on crop management methods that protect soils instead. These may include selecting crops that help to protect soils (bush or tree crops) or crops that cause less disturbance to soils. Examples of interventions for protecting against soil degradation include avoiding root crops, using mulches to anchor and stabilize soils, and planting crops on river banks that can tolerate intermittent flooding.

Soil fertility

UA is generally more intensive than agriculture in rural areas. The predominance of tree crops and short-cycle horticulture crops in UA is expected to contribute towards high nutrient demand from soils. The balance between farm gate nutrient imports and exports is essential for the long-term sustainability of any agricultural system. Fortunately, the predominance of organic wastes, rich in nitrogen (N) and phosphorus (P), provides UA with a steady supply of nutrients, and in some cases can lead to significant surpluses (Khai et al., 2007; Wang et al., 2008). Although nutrient surpluses can cause significant environmental problems to urban waterways, they tend not to be a constraint to crop production. However, the over-reliance on a limited variety of organic wastes as the sole source of fertilizer can lead to soil nutrient deficiencies if the waste contains insufficient amounts of certain micronutrients. Depending on the source material, organic wastes are likely to provide sufficient or even an excess of N and P but may provide insufficient potassium (K). Wang et al. (2008) reported that the continued use of organic fertilizers led to negative soil K balances in peri-urban farms in China. This emphasizes the need for farmers to develop balanced nutrient management plans even for plot-scale farming: poorer and less educated farmers have less access to inorganic and animal manure fertilizers and are less likely to be concerned about distributing evenly the nutrients they can access (Zingore et al., 2007).

Soil pollution

The pollution of urban soils can also have negative impacts on crops. A major source of soil pollution is from the use of urban wastes for soil amendments (Alloway, 1995). These urban wastes, such as sewage sludge, black soils, or municipal solid wastes (MSW), are mostly partially decomposed organic wastes and therefore make excellent soil amendments that are rich in nutrients and organic matter. In addition, the low cost and high availability of these wastes make their use in the UA of southern cities very common (Eaton and Hilhorst, 2003). Another source of soil

pollution is from the burning of wastes or incomplete combustion of fossil fuels from which heavy metals and other pollutants are deposited by air (Alloway, 1995; Elaigawu et al., 2007).

There is evidence to show that the long-term burning of municipal wastes causes an increase in heavy metal concentration and soil salinity that can negatively affect crop production (Anikwe and Nwobodo, 2002; Elaigawu et al., 2007; Hargreaves et al., 2007). Another example of the impact of pollution in urban areas (Aubry et al., 2008) arises from the uncontrolled release of industrial liquid wastes into irrigation networks in Antananarivo, Madagascar; these negatively affected rice production and have led to a reduction in urban rice growing. However, the impact of urban wastes on crop yields is not consistent across crops and places, and is expected to depend on the duration of application, the pollutant concentration in the waste, and the fertilizer value of the waste (Hargreaves et al., 2007). Most research has been conducted on clean soils which do not account for the numerous and repeated applications typically occurring in the UA of southern cities. It is precisely the build-up of heavy metals to certain significant levels that can cause negative impacts such as a decline in crop yields.

A good intervention would be to equip farmers better with the knowledge and skills to identify and evaluate the quality of the wastes being used. With an improved capacity and understanding of wastes, farmers can monitor and amend the frequency of application and the quantity of wastes applied. Avoiding any risk of contamination requires a safe substitute soil amendment being available at similar cost; otherwise, farmers will continue to use wastes even in the face of health and biosafety legislation. There is also the possibility of intervention by city governments or private companies to turn waste into valuable soil ameliorants: fly ash, a common urban soil contaminant in southern cities, could be a valuable soil ameliorant, improving water-holding capacity and nutrient status (Jala and Goyal, 2006).

Water availability and soil drainage

Both urban and rural agriculture are dependent on the same seasonal rainfall patterns, runoff, and irrigation from remote sources, e.g. rivers or systems for storage and distribution primarily for human consumption, Runoff, with unpredictable occurrence and possibly damaging volumes, is more significant in UA because of the preponderance of hard surfaces in cities. Water from sewage and domestic wastewater is also important, although authorities do not generally support the use of treated water for irrigated agriculture because of the challenges of meeting the demand for urban dwellers. When urban farmers are willing to pay for the full cost of water supply or to pay at the same rate as other users, city authorities provide treated water for irrigation (e.g. in Lomé, Togo and Accra, Ghana). In most urban areas, farmers have no access to water for irrigation and have to rely on wastewater. As the rate of urbanization increases in southern cities and investment in water supply far outpaces that of sanitation and waste management, competing uses of wastewater for irrigated vegetables, fodder, ornamental and other crops will continuously expand. In general, the use of wastewater provides benefits to UA but has added risks to both farmers and consumers. Scott *et al.* (2004) identifies the main benefits of wastewater use as availability and lower costs, but it can cause chronic diarrhoea and gastrointestinal diseases as negative outcomes. Wastewater irrigation of vegetables in urban areas may serve as a transmission route for heavy metals (Scott *et al.*, 2004). Restricting the use of wastewater is not an option to millions of urban farmers in the southern cities whose livelihoods depend on UA (Scott *et al.*, 2004; Obuobie *et al.*, 2006).

The demand for water can be a constraint in UA when water-intensive horticultural crops are widely planted in urban areas without sufficient water supply. Ashebir *et al.* (2007) reported that, as urban farmers in Ethiopia shift towards more water-intensive horticultural crops, the demand for water may exceed supply and become a major production constraint. For example, typical UA crops such as lettuce, cabbage and spring onions need to be irrigated twice per day (Obuobie *et al.*, 2006). Apart from access to water in urban areas, there are fewer opportunities for farmers in terms of water-lifting technologies.

A further water-related constraint is water drainage. UA in low-lying lands such as flood-prone valley bottoms or river banks may experience conditions of excess water. Poor drainage, ponding or flooding of agricultural lands can lead to significant loss of crops, nutrient leaching and greater incidence of soil-borne diseases. Agricultural drainage requires a significant investment of capital, material and human resources, but the uncertain land tenure of most UA systems discourages these investments. Nevertheless, some low-cost interventions are available to farmers such as raised beds, furrows, or planting of crops well adapted to growing in water, such as rice or taro. Farmers' responses to UA water issues may include flexible sourcing of water, e.g. from household wastewater in dry periods, drainage and irrigation. Other management adaptations include mulching the soil, which increases water-holding capacity and thus retention of rainwater (Ramakrishna *et al.*, 2006).

Pests

There has been little study of the incidence or impact of plant pests and diseases in UA. The establishment, persistence and impact of plant diseases are generally dependent on crop diversity (mixed cropping), rotations, and other pest management practices (such as burning diseased plant residues), lack of money for synthetic pesticides, and environment, e.g. high humidity, lower irradiance and difference spectra, and lack of air movement in cities. There is no reason for these cultural and management practices to be different between UA and rural agriculture, though urban farmers may have greater labour constraints. A case study of urban horticulture in Lomé, Togo identified three major constraints faced by farmers to control pests: difficulty in identifying pests, high costs of pesticides and resistance to pesticides (Tallaki, 2005). The former is partly due to a lack of agronomic extension services in urban areas. The recognition by municipal planners of the importance of urban extension services has been slow, but where action has taken place (e.g. Mekelle, Ethiopia) these services have been well received (Ashebir *et al.*, 2007). The sizes of fields in UA are generally much smaller than in rural areas, and support an equal or greater variety of crops. A greater variety of crops planted on smaller, less contiguous and more dispersed fields is a recognized crop management method that prevents the spread and establishment of pests and diseases (Lampkin, 1990). The diversity of crop type and range of field sizes in UA may also be of benefit in preventing the spread of diseases by vectors.

Theft or damage of crops by humans is a common problem faced by urban farmers (Bryld, 2003). In some cities in Africa it is a considerable problem, with over half of urban farmers experiencing theft on a daily basis (Freeman, 1993; Smith, 1996). In Kinshasa, the loss of vegetables at night was reported as a common (Mayeko, 2009). Humans trespassing through fields are the most common cause; however, motor vehicles and animals can cause significant damage to crops if they are close to roads or busy thoroughfares. The susceptibility of UA to theft is partly due to it being practised on common lands where access is open and legal ownership of the crops is tenuous. Improved fencing or increased vigilance by watchmen can reduce theft. Successful adaptations that reduce the likelihood of theft include harvesting crops earlier and planting buffers strips with low quality crops or non-food plants. Boudjenouia *et al.* (2008) report that urban farmers in Algeria prefer to plant cereal crops instead of higher value leguminous crops on lands where the risk of theft is high. Ideally, improved community security, neighbourhood watch and public awareness should be promoted as solutions. Planners and policy makers with responsibility for community planning and security should at least be made aware of the risk of human theft when identifying areas for UA. In addition, interventions that strengthen and improve the capacity of farmer organizations should be promoted.

Conclusions

This review has identified unique constraints. The extent and impact that these have on crop yield varies considerably. Most agronomic constraints act at the plant and field level and require interventions by farmers. On the other hand, a number of interventions require actions at the broader city or regional level. At the field level, recommended interventions should be assessed by farmers as to their potential to increase

yields and as to practical considerations of the required cost and resources of the intervention. Significant progress by farmers in overcoming these agronomic constraints will be likely to require strengthened capacity of farmers and farmer organizations to access training and extension services, as well as technology and tools. At the broader city or regional level, these constraints should be addressed by policy makers, urban planners and urban agricultural extension agents at higher policy and planning levels. In addition, to ensure the sustainability of food production and food markets within a city, it is advisable that farmers are not only consulted but also included in any planning and policy decisions that might impact their farms. The food security of the urban poor will therefore require coordinated efforts and cooperation between the farmers who produce food, and planners and policy makers who manage the supporting systems such as markets, inputs and land registration. Furthermore, this coordinated effort should focus on exploring different funding opportunities from the city authorities and international sources for the unique research gaps identified in this paper. Policy makers should also provide resources to support the development of UA in cities.

References

Afrane, Y. A., Klinkenberg, E., Drechsel, P., Daaku, K., Garms, R., Kruppa, T., 2004, 'Does irrigated urban agriculture influence the transmission of malaria in the city of Kumasi, Ghana?', *Acta Tropica. Malaria and Agriculture* 89, 125–134.

Ainsworth, E. A., Long, S. P., 2005, 'What have we learned from 15 years of free-air CO_2 enrichment (FACE)?', *New Phytologist* 165, 351–371.

Alloway, B. J., 1995, *Heavy Metals in Soils* (2nd edn.), Blackie, London.

Alpert, P., Kishcha, P., 2008, 'Quantification of the effect of urbanization on solar dimming', *Geophysical Research Letters* 35, L08801.

Anikwe, M. A. N., Nwobodo, K. C. A., 2002, 'Long term effect of municipal waste disposal on soil properties and productivity of sites used for urban agriculture in Abakaliki, Nigeria', *Bioresource Technology* 83, 241–250.

Ashebir, D., Pasquini, M., Bihon, W., 2007, 'Urban agriculture in Mekelle, Tigray state, Ethiopia: principal characteristics, opportunities and constraints for further research and development', *Cities* 24, 218–228.

Aubry, C., Ramamonjisoa, J., Dabat, M. H., Rakotoarisoa, J., Rakotondraibe, J., Rabeharisoa, L., 2008, 'L'agriculture à Antananarivo (Madagascar): une approche interdisciplinaire', *Natures Sciences Sociétés* 16, 23–35.

Bhatt, V., 2005, 'Project background', in: V. Bhatt, R. Kongshaug (eds), *Making the Edible Landscape EL. A Study of Urban Agriculture in Montreal*, School of Architecture, McGill University, Montreal.

Boa, W., Zhu, Z., 1997, *Agro-Environmental Protection* [in Chinese] 16, 16.

Boudjenouia, A., Fleury, A., Tacherift, A., 2008, 'L'agriculture périurbaine à Sétif (Algérie) quel avenir face à la croissance urbaine?', *Biotechnology, Agronomy, Society and Environment* 12, 23–30.

Bryld, E., 2003, 'Potentials, problems and policy implications for urban agriculture in developing countries', *Agriculture and Human Values*, 20 (1), 79–86.

Cabannes, Y., Dubbeling, M., 2001, 'Urban agriculture and urban planning: what should be taken into consideration to plan the city of the 21st century', in: H. Hoffmann, K. Mathey (eds), *Urban Agriculture and Horticulture, the Linkage with Urban Planning*, International Symposium, Berlin, July.

Cohen, S., Raveh, E., Li, Y., Grava, A., Goldschmidt, E. E., 2005, 'Physiological responses of leaves, tree growth and fruit yield of grapefruit trees under reflective shade screens', *Scientia Horticulturae* 107, 25–35.

Douglas, I., Alam, K., Maghenda, M., McDonnell, Y., McLean, L., Campbell, J., 2008, 'Unjust waters: climate change, flooding and the urban poor in Africa', *Environment and Urbanization* 20, 187–205.

Drechsel, P., Cofie, O., Fink, M., Danso, G., Zakari, F. M., Vasquez, R., 2004, *Closing the Rural–Urban Nutrient Cycle. Options for Municipal Waste Composting in Ghana*, Final Scientific Report submitted to IDRC (Project 100376).

Eaton, D., Hilhorst, T., 2003, 'Opportunities for managing solid waste flows in the peri-urban interface of Bamako and Ouagadougou', *Environment and Urbanization* 15, 53–63.

Elaigawu, S. E., Ajibola, V. O., Folaranmi, F. N., 2007, 'Studies on the impact of municipal waste dumps on surrounding soil and air quality of two cities of Northern Nigeria', *Journal of Applied Sciences* 7, 421–425.

Emberson, L. D., Ashmore, M. R., Murray, F., Kuylienstierna, J. C. I., Percy, K. E., Izuta, T., Zheng, Z., Shimizu, H., Sheu, B. H., Liu, C. P., Agrawal, M., Wahid, A., Abdel-Latif, N. M., van Tienhoven, M., de Bauer, L. I., Domingos, M., 2001, 'Impacts of air pollution on vegetation in developing countries', *Water, Air, and Soil Pollution* 130, 107–118.

Foeken, D., Mwangi, A. M., 2000, 'Increasing food security through urban farming in Nairobi', in: N. Bakker, M. Dubbeling, S. Gundell, U. Sabel-Koschella, H. de Zeeuw (eds), *Growing Cities, Growing Food: Urban Agriculture on the Policy Agenda: A Reader on Urban Agriculture*, Food and Agriculture Development Centre, Feldafing, Germany, 303–328.

Freeman, D. B., 1993, 'Survival strategy or business training ground? The significance of urban agriculture for the advancement of women in African cities', *African Studies Review* 36, 1–22.

Garnett, T., 1996, *Growing Food in Cities. A Report to Highlight and Promote the Benefits of Urban Agriculture in the UK*, National Food Alliance and SAFE Alliance.

Hargreaves, J. C., Adl, M. S., Warman, P. R., 2007, 'A review of the use of composted solid waste in agriculture', *Agriculture, Ecosystems & Environment* 123, 1–14.

Jacobi, P., Amend, J., Kiango, S., 2000, 'UA in Dar es Salaam: providing for an indispensable part of the diet', in: N. Bakker, M. Dubbeling, S. Gundel, U. Sabel-Koschella, H. de Zeeuw (eds), *Growing Cities, Growing Food: Urban Agriculture on Policy Agenda: A Reader on Urban Agriculture*, Food and Agriculture Development Centre, Feldafing, Germany, 257–283.

Jala, S., Goyal, D., 2006, 'Fly ash as a soil ameliorant for improving crop production: a review', *Bioresource Technology* 97, 1136–1147.

Jamal, V., 1988, 'Coping under crisis in Uganda', *International Labour Review* 127, 679–701.

Khai, N. M., Ha, P. Q., Oborn, I., 2007, 'Nutrient flows in small-scale peri-urban vegetable farming systems in Southeast Asia : a case study in Hanoi', *Agriculture, Ecosystems & Environment* 122, 192–202.

Krupa, S., McGrath, M. T., Andersen, C. P., Booker, F. L., Burkey, K. O, Chappelka, A. H., Chevone, B. I., Pell, E. J., Zilinskas, B. A., 2001, 'Ambient ozone and plant health', *Plant Disease* 85, 4–12.

Kumar, D., Tieszen, L. L., 1980, 'Photosynthesis in Coffea Arabica. I. Effects of light and temperature', *Experimental Agriculture* 16, 13–19.

Lampkin, N., 1990, *Organic Farming*, Farming Press, Miller Freeman House, Tonbridge, UK.

Liang, W., Wang, M., Xizhen, Ai, 2009, 'The role of calcium in regulating photosynthesis and related physiological indexes of cucumber seedlings under low light intensity and suboptimal temperature stress', *Scientia Horticulturae* 123, 34–38.

Mayeko, K. K., 2009, 'Wastewater use and urban agriculture in Kinshasa, DR Congo', in: M. Redwood (ed.), *Agriculture in Urban Planning: Generating Livelihoods and Food Security*, Earthscan/IDRC, London, 147–166.

Maxwell, D. G., 1994, 'The household logic of urban farming in Kampala', *Cities Feeding People: an Examination of Urban Agriculture in East Africa*, IDRC, Ottawa, Canada, 47–66.

Maxwell, D. G., Zziwa, S., 1992, *Urban Agriculture in Africa: The Case of Kampala, Uganda*, African Centre for Technology Studies, Nairobi, Kenya.

Mbaye, A., Moustier, P., 2000, 'Market-oriented urban agricultural production in Dakar', in: N. Bakker, M. Dubbeling, S. Gundel, U. Sabel-Koschella, H. de Zeeuw (eds), *Growing Cities, Growing Food: Urban Agriculture on Policy Agenda: A Reader on Urban Agriculture*, Food and Agriculture Development Centre, Feldafing, Germany, 235–256.

Mougeot, L. J. A., 2006, *Growing Better Cities: Urban Agriculture for Sustainable Development*, IDRC, Ottawa, Canada.

Mourão, I. M. G., Hadley, P., 1998, 'Environmental control of plant growth development and yield in broccoli (*brassica oleracea* I. Var. *Italica* plenk): crop responses to light regime', *Acta Horticulturae* 459, 71–78.

Nabulo, G., Oryem-Origa, H., Diamond, M., 2006, 'Assessment of lead, cadmium and zinc contamination of roadside soils, surface films, and vegetables in Kampala City, Uganda', *Environmental Research* 101, 42–52.

Nsangu, C. A., 2008, 'Urban agriculture and physical planning: a case study of Zaria, Nigeria', in: M. Redwood (ed.), *Agriculture in Urban Planning: Generating Livelihoods and Food Security*, Earthscan/IDRC, London, 217–234.

Obuobie, E., Keraita, B., Danso, G., Amoah, P., Cofie, O., Raschid-Sally, L., Drechsel, P., 2006, *Irrigated Urban Vegetable Production in Ghana: Characteristics, Benefits and Risks*, IWMI-RUAF-CPWF, Accra, Ghana.

Ramakrishna, A., Hoang Minh, T., Sukas, W., Tranh Dinh, L.., 2006, 'Effect of mulch on soil temperature, moisture, weed infestation and yield of groundnut in northern Vietnam', *Field Crops Research* 95, 115–25.

Rich, S., 1964, 'Ozone damage to plants', *Annual Review of Phytopathology* 2, 253–266.

Rodriguez-Flores, M., Rodriguez-Castellon, E., 1982, 'Lead and cadmium levels in soil and plants near highways and their correlation with traffic density', *Environmental Pollution B* 4, 281–290.

RUAF (Resource Centres on Urban Agriculture and Food Security), 2007, 'Villa Maria del Truinfo (Lima, Peru)' [available at www.ruaf.org/node/509].

Sanchez, C. A., Allen, R. J., Schaffer, B., 1989, 'Growth of yield of crisphead lettuce under various shade conditions', *Journal of the American Society for Horticultural Science* 114, 884–890.

Sawio, C. J., 1993, 'Feeding the urban masses?', *Towards an Understanding of the Dynamics of Urban Agriculture and Land Use Change in Dar es Salaam, Tanzania*, PhD thesis, Clark University, Worcester, MA.

Scott, C., Faruqui, N. I., Raschid-Sally, L., 2004, *Wastewater Use in Irrigated Agriculture Confronting the Livelihood and Environmental Realities*, CABI/IWMI/IDRC, London.

Smith, D., 1996, 'Urban agriculture in Harare: socio-economic dimensions of a survival strategy', in: D. Grossman, L. M. van den Berg, H. I. Ajaegbu (eds), *Urban and Peri-urban Agriculture in Africa, Proceedings of a Workshop, Netanya, Israel, 23–27 June*, Ashgate, Aldershot, UK.

Stanhill, G., Cohen, S., 2001, 'Global dimming: a review of the evidence for a widespread and significant reduction in global radiation with discussion of its probable causes and possible agricultural consequences', *Agricultural and Forest Meteorology* 107, 255–278.

Tallaki, K., 2005, 'The pest-control system in market gardens of Lomé, Togo', in: L. Mougeot (ed.), *Agropolis: The Social, Political, and Environmental Dimensions of Urban Agriculture*, Earthscan, London, 50–87.

Taylor, J., Paine, C., FitzGibbon, J., 1995, 'From greenbelt to greenways: four Canadian case studies', *Landscape and Urban Planning* 33, 47–64.

Torres-Lima, P., Rodriguez Sanchez, L. M., Garcia Uriza, B. I., 2000, 'Mexico City: The integration of urban agriculture to contain urban sprawl', in: N. Bakker, M. Dubbeling, S. Gundel, U. Sabel-Koschella, H. de Zeeuw (eds), *Growing Cities, Growing Food: Urban Agriculture on Policy Agenda: A Reader on Urban Agriculture*, Food and Agriculture Development Centre, Feldafing, Germany, 363–390.

Van Ittersum, M. K., Rabbinge, R., 1997, 'Concepts in production ecology for analysis and quantification of agricultural input–output combinations', *Field Crops Research* 52, 197–208.

Van Veenhuizen, R., Danso, G., 2007, 'Profitability and sustainability of urban and peri-urban agriculture', in:

Agricultural Management, Marketing and Finance, Occasional Paper No. 19, FAO, Rome.

Vogl, C. R., Axmann, P., Vogl-Lukasser, B., 2004, 'Urban organic farming in Austria with the concept of *Selbsternte* ('self-harvest'): an agronomic and socio-economic analysis', *Renewable Agriculture and Food Systems* 19, 67–79.

Wade, I., 1987, *Food Self-Reliance in Third World Cities,* The Food-Energy Nexus Programme, UN University, Paris.

Wang, G. G., Qian, H., Klinka, K., 1994, 'Growth of *Thuja plicata* seedlings along a light gradient', *Canadian Journal of Botany* 72, 1749–1757.

Wang, H.-J., Huang, B., Shi, X.-Z., Darilek, J. L., Yu, D.-S., Sun, W.-X., Zhao, Y.-C., Chang, Q., Oborn, I., 2008, 'Major nutrient balances in small-scale vegetable farming systems in peri-urban areas in China', *Nutrient Cycling in Agroecosystems* 81, 203–218.

Ward, N. I., Reeves, R. D., Brooks, R. R., 1975, 'Lead in soil and vegetation along a New Zealand State highway with low traffic volume', *Environmental Pollution* 9, 243–251.

Yeung, Y. M., 1985, *Urban Agriculture in Asia*, Food-Energy Nexus Programme, UN University, Tokyo, Japan.

Zheng, Y., Chen, S., 1991, *Atmospheric Environment (China)* [In Chinese] 6, 45.

Zingore, S., Murwira, H. I. C., Delve, R. J., Gillert, K. E., 2007, 'Influence of nutrient management strategies on variability of soil fertility, crop yields and nutrient balances on smallholder farms in Zimbabwe', *Agriculture, Ecosystems and Environment* 119, 112–126.

Urban agriculture and sanitation services in Accra, Ghana: the overlooked contribution

Mary Lydecker[1]* and Pay Drechsel[2]

[1] Harvard University Graduate School of Design, 26 Dimick Street, Somerville, MA 02143, USA
[2] International Water Management Institute (IWMI) and Resource Centers for Urban Agriculture and Food Security (RUAF), PMB CT 112, Accra, Ghana

While urban agriculture has long been valued for providing food security and nutrition within cities, it contributes to many other urban services that are seldom cited as rationales for protecting or even expanding urban food production. Articulating the actual and possible contributions of urban agriculture to municipal sanitation and health services is critical for sustaining these urban farms and their functions into the future. In the context of the low coverage and performance of wastewater treatment plants in Accra, Ghana, health risk reduction measures implemented on and off farm can substitute to a large extent for this absence of conventional wastewater treatment. We estimate that Accra generates approximately 80,000,000L of wastewater per day, of which urban vegetable farms alone use up to 11,250,000L. By mitigating the health risks for farmers and consumers associated with widespread wastewater irrigation, these urban farms have the potential to significantly contribute to the city's sanitation needs. This could allow partial outsourcing of public health services from treatment plants to the farm, where wastewater is considered an asset instead of a problem. Urban agriculture could also significantly support buffer zone management along streams and rivers, resulting in a reduction of solid waste dumping and environmental pollution, but most importantly an improvement in flood control and related public health challenges. While urban agriculture is not the panacea for addressing these urban challenges, it can significantly contribute to their solution.

Keywords: environment, health, sanitation, services, urban planning, water management

Introduction

Many of the urban agriculture plots in Accra have been cultivated for decades. However, almost all farmers of these lands practise without secure tenure; farming is based on informal agreements with private or governmental landowners (Flynn-Dapaah, 2002). As development increases at a rapid rate within the city, the sustainability of these urban farms is under discussion (Drechsel and Dongus, 2009). The future of urban agriculture in Accra and other cities depends on a critical evaluation of all the urban services that it already provides and of how these services may be augmented through strategic planning, using urban agriculture to frame and serve urban development. In addition to the contributions of urban agriculture to urban food security, nutrition and local economies, farming also contributes to urban water management, sanitation and health services. Since urban farmers in Accra have little alternative but to use wastewater to irrigate their crops, sustaining urban agriculture in Accra requires a reduction in health risks associated with this practice; in addition to augmenting the long-term value and existence of farms in Accra, decreasing these risks will improve urban sanitation by safeguarding public health.

The contribution of urban agriculture to municipal sanitation is evaluated in this paper, based on its direct and indirect contributions to urban sanitation and development goals.

Limitations and opportunities for policy support

Urban crop production in Ghana has two common faces: backyard farming primarily for personal consumption

*Corresponding author. Email: mary.lydecker@gmail.com

INTERNATIONAL JOURNAL OF AGRICULTURAL SUSTAINABILITY 8 (1&2) 2010

PAGES 94–103, doi:10.3763/ijas.2009.0453 © 2010 Earthscan. ISSN: 1473-5903 (print), 1747-762X (online). www.earthscan.co.uk/journals/ijas

and market-oriented open-space farming on larger plots. While backyard gardens are in general considered as private concerns, open-space farming occupies significant public areas, though it usually falls in the informal sector. For instance, the Ghana Irrigation Development Authority (GIDA), which is the government agency officially responsible for developing irrigation in the country, has always focused solely on public irrigation schemes in rural areas and for many years considered irrigated vegetable farming in urban or peri-urban areas as outside its jurisdiction. Only the new National Irrigation Policy and the recent decentralization of Ghana's Ministry of Food and Agriculture gave the sector more attention (Obuobie *et al.*, 2006).

Despite this official acknowledgement, the integration of urban agriculture into sustainable urban development is another step. In 2005, the network of the Resource Centers for Urban Agriculture and Food Security (RUAF) took up this challenge by initiating multi-stakeholder (MS) processes in 20 cities of 17 countries in Asia, Latin America and Africa, including Accra. RUAF tried to facilitate strategic partnerships for an improved research-policy dialogue. The process was supported by capacity building of local stakeholders, e.g. in participatory processes management, action planning and research, and monitoring and evaluation. One key lesson learned was that there are diverse reasons why local partners might not give every project the expected priority and commitment (Drechsel *et al.*, 2008).

To increase and better understand local commitment, Obuobie *et al.* (2006) recommended an analysis beforehand of: (a) institutional and individual priorities, constraints and capacities of potential partners, and (b) how urban agriculture links to the partners' institutional strategies and work plans and how these could benefit from the MS process and urban and peri-urban agriculture. In this way it might be possible to avoid the impression that something 'external' will be added to an already full work plan (Obuobie *et al.*, 2006). In other words, the theoretical advantages of urban and peri-urban agriculture – such as contribution to food supply, livelihoods and urban greening or training in a new subject – need to be made explicit and clearly linked to existing work plans of the involved institutions. It will be crucial to highlight, for each partner, the actual and possible added value of – for example – irrigated urban agriculture to urban flood control, buffer zone management, environmental protection, biodiversity or wastewater treatment, i.e. issues already on their agenda (Drechsel *et al.*, 2008).

Sanitation sector on the verge of collapse

In this paper, we discuss some of the possible linkages between open-space urban agriculture and solid and liquid waste management in Ghana. Waste management is certainly one of the greatest challenges and priority issues in most cities, especially in the developing world (Satterthwaite and McGranahan, 2007). Ghana's cities are no exception.

Ghana's capital city Accra is growing in leaps and bounds; so also is the generation of waste. Currently, the population of the city and its neighbouring metropolitan areas is around 3 million people with a floating population of over 500,000 a day. These together generate about 1800 tons of garbage daily, which need to be evacuated for final disposal. Out of this amount, approximately 1200–1300 tons is collected daily, leaving a backlog of between 500–600 tons. The huge backlog is reflected in choked drains, overflowing garbage heaps, littered pavements and polluted streams (AMA, 2006).

The state of wastewater and faecal sludge management in urban Ghana is not unlike many other parts of the developing world: the supply of human waste far outstrips the existing capacity of treatment facilities. This situation is mirrored in the solid waste sector and other urban services as sanitation fees and tax revenues are generally so low that cost recovery is not even in the vocabulary of Accra's Revenue Mobilization, Sanitation and Waste Management Strategy. Indeed, environmental sanitation currently absorbs an average of 35 per cent of the municipal budget, making the city dependent on governmental support (Government of Ghana, 2008). However, even this allocation is neither sufficient to maintain the system nor to keep pace with urbanization.

A nationwide assessment of the state of wastewater and faecal sludge treatment plants in Ghana showed that of the 70 identified plants fewer than 10 per cent were operating as designed (IWMI, 2009). More than 85 per cent of wastewater and faecal sludge that is generated every day is discharged into seldom desludged and usually leaking septic tanks or straight into the environment without any effective collection, treatment or disposal.

Accra hosts about half of Ghana's wastewater and sludge treatment plants. Most serve smaller communities or public institutions, like the local university, military camp or major hospitals. Thus the sewer system is largely decentralized. An exception is the larger James Town UASB (Upflow Anaerobic Sludge

Blanket) treatment plant at the Korle Lagoon, down-town Accra, which serves the central administrative district of the city. UASB technology, normally referred to as a UASB reactor, is a form of anaerobic digester that is used in the conventional treatment of wastewater and considered a robust technology for developing countries.

The James Town UASB sewage treatment plant was constructed in Accra in 2000, with a design capacity of 16,120,000 litres of wastewater per day; at a per capita rate of 50 litres of wastewater per day, this potentially serves 322,400 people living or working in adjacent neighbourhoods (Awuah and Abrokwa, 2008[1]). However, the plant has not functioned since 2004, due to broken pumps, lack of funds and institutional capacity challenges after the decentralization of the sanitation sector. Today, it is little more than a convergence point for untreated wastewater. While the plant had a designed emergency ocean outfall in the Gulf of Guinea, this outlet pipe also broke, sending untreated water directly into the Korle Lagoon. Independently of this development, the lagoon is undergoing a major ecological restoration project, which is undermined by the Lagoon inflow from the UASB plant exactly in its newly protected zone. The failure of the UASB might be attributed to the general challenges of centralized treatment plants (Nhapi and Gijzen, 2004). However, as most decentralized plants in Ghana are also in critical need of rehabilitation or replacement, the problem appears to go deeper. Until public services can maintain their systems to effectively and consistently serve the city, other options for safeguarding environmental and public health need to be explored in order to improve municipal sanitation at least in the short to medium term. Indeed, if functioning properly, Accra's sewerage systems and wastewater treatment plants only serve 5–7 per cent of Accra's population (Obuobie et al., 2006). The large majority (90 per cent) of the collected faecal sludge is currently dumped into the ocean. Evaluating alternative treatment methods and protecting the public from irrigation with untreated water are critical challenges to which urban agriculture can contribute.

Irrigated urban agriculture and public health

In Accra, about 680ha of urban land are under maize cultivation, 47ha under vegetables and 251ha under mixed cereal-vegetable systems. Irrigated urban vegetable production takes place on approximately 75ha,

divided among seven large sites across the city (Obuobie et al., 2006). Recent construction work is reducing one of the traditional sites, while in other parts of the city new farms appear – a dynamic that is typical for urban agriculture (Drechsel and Dongus, 2009). All plots used for vegetable farming are close to streams and stormwater drains, since the most profitable exotic vegetables require continuous irrigation. The availability of safe water, however, even outside the city, is poor given the small coverage and condition of wastewater treatment plants. This situation is representative for the whole sub-region, posing a critical health risk to farmers and urban vegetable consumers (Drechsel et al., 2006).

The insecure tenure of urban farmers in Accra contributes to the reluctance of farmers to invest in access to safer water through on-site pumps or piped sources (Flynn-Dapaah, 2002). Irrigation relies on manual fetching or pumps and then takes place with watering cans. On other sites, normal waterways are temporarily blocked to channel the water into furrows. This system actually allows a productive use of wastewater. It capitalizes on the nutrients in the water, which conventional treatment would remove without benefit, and simultaneously forms a decentralized land application system with wastewater filtration through the soil. As Ghana's urban wastewater in general does not contain any particular chemical threats but is rich in pathogens from faecal matter, land application can actually be more effective than what currently happens in most of the country's treatment plants. The same applies to faecal sludge. Farmers, especially in the north but also in Accra, ask drivers of septic trucks to desludge on or near their fields instead of at an official site. The liquid sludge might be used directly as fertilizer on maize (Accra) or the sludge is allowed to dry and then spread on sorghum fields. Health risks for consumers are marginal (Cofie et al., 2008), while the profits allow farmers to buy anti-helminths ('dewormer') if needed.

Farmers are well aware of the fertilizer value of the sludge, which resembles cow manure when dried. However, while nobody rings an alarm bell when farmers use farmyard manure, the use of human excreta or wastewater is a thorn in the public side. The health risks from sludge, excreta and wastewater for crop cultivation have been carefully studied in Ghana in recent years, and the risk from wastewater use for consumers of raw eaten vegetables are indeed significant. To address these risks, a number of risk mitigation approaches and practices have been developed to control them (Drechsel et al., 2008, 2010). Some of these are shown in these online awareness and training videos:

- http://video.google.com/videoplay?docid=-3530336707586348166&hl=en (on good farming practices)
- http://video.google.com/videoplay?docid=-6891955003003280662&hl=en (on good practices for street food restaurants)

The principal concerns include gastrointestinal diseases. Songsore *et al.* (2005) rank gastrointestinal diseases as the second most important health problem after malaria for most communities in Accra, especially in high-density, low-income areas. Diarrhoea is among the top five reported causes of morbidity in Ghana for people of all ages (Ghana Ministry of Health, 2007). Aside from poor sanitation and hygiene, food safety is also a contributing factor as polluted water is used for irrigation in and around the cities. A detailed microbiological risk assessment estimated about 12,000 lost healthy life years (DALYS, disability adjusted life years) annually in Ghana's major cities through the consumption of salad prepared from wastewater-irrigated lettuce. This figure represents nearly 10 per cent of the World Health Organization's reported losses due to various types of water and sanitation related cases of diarrhoea in urban Ghana (Ghana Statistical Service, 2004; Prüss-Üstün *et al.*, 2008). Compared with contaminated drinking water, flooding, risks from open drains, swimming at urban beaches, and occupational contact with faecal matter, the consumption of wastewater irrigated vegetables ranks second to contact with open drains in terms of risks to urban dwellers (IWMI, 2009).

These quantitative microbiological risk assessments demonstrate that of the above-mentioned 12,000 lost healthy life years from eating wastewater irrigated salads, more than 90 per cent can be avoided through interventions at the farm and/or post-harvest sectors (IWMI, 2009). This shows that it is possible to protect public health from wastewater irrigation even in situations where conventional treatment mechanisms fail. We do not want to advocate a situation without conventional treatment. However, we intend to stress the advantages of urban sanitation management that emphasizes reuse as well as complementary risk reduction strategies, e.g. on farms; the World Health Organization's (2006) guidelines for safe wastewater irrigation recommend such a multiple barrier approach to sanitation.

Two major opportunities for improving urban sanitation through urban agriculture exist in Accra. The first opportunity includes the creation or rehabilitation of wastewater treatment plants for targeted reuse; for example, non- or partially treated water could be diverted for agriculture located immediately around the plant. IWMI is currently proposing with the Government of Ghana a pilot project that will test the viability of this method, as an alternative to the existing 'treatment for disposal' approach. The second opportunity includes on-farm and post-harvest water treatment options that can significantly reduce the health risks associated with wastewater-irrigated crops (Figure 1). The benefits of this approach include its immediate applicability to the current situation at sites in Accra and the existence of materials and network of governmental (farmer extension programme) and non-governmental (IWMI and RUAF extension support materials) resources for short- to medium-term implementation.

At La Fulani urban agriculture site, a combination of the above two opportunities is already being practised. Wastewater from the Burma Military barracks' non-functioning wastewater treatment system is diverted by farmers into a furrow irrigation system for maize

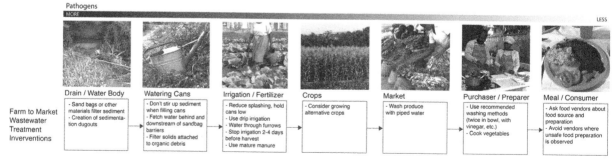

Figure 1 | Farm to market wastewater treatment interventions
Source: IWMI educational outreach video, 2008.

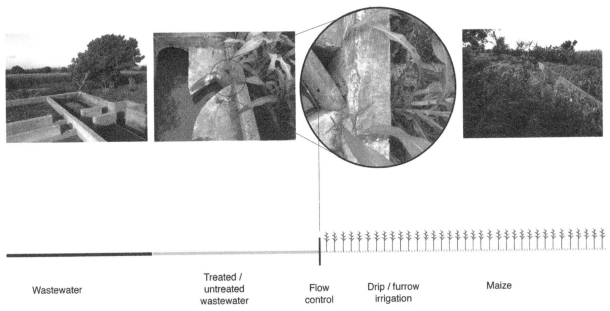

Wastewater   Treated /
untreated
wastewater   Flow
control   Drip / furrow
irrigation   Maize

Figure 2 | The flow of wastewater from Burma Camp's broken down treatment system is redirected and used for maize production. This is one example of how urban agriculture in Accra may be integrated with conventional wastewater treatment systems to maximize the reuse potential of wastewater.

crops (Figure 2). Outflow from the final settlement pond is also blocked with sandbags by farmers to create a sequence of labour-saving fetching ponds for irrigation of vegetables with watering cans. These informal sedimentation ponds allow pathogens, especially helminth eggs to settle; the efficiency of this settling depends on the retention time of the water. Pathogen inactivation from UV radiation also occurs, based on the exposure of each pond to direct sunlight. IWMI is planning further studies on the extent of pathogen inactivation and removal in these informal settlement pond systems. The quality of irrigation water that is ultimately used from these ponds then depends on lower-risk irrigation practices of farmers as illustrated above. Urban agriculture at the La Fulani site can be seen in the following online video:

- http://video.google.com/
  videoplay?docid=-788126851657143043&hl=en

Pollution reduction

As stressed above, conventional wastewater treatment remains critical. Urban farms can only absorb and filter a certain amount of the generated urban wastewater, thus the majority of all wastewater still pollutes

the environment. But at least the part in contact with the food chain can be treated and the risks controlled.

As the weather is hot in Ghana, farms have to be irrigated twice a day, unless it rains. Vegetable farmers in Accra use 11,250,000 litres of irrigation water per day; the majority of this water is urban grey water, raw or mixed with stream or river water, which is often diverted into shallow standing-water dugouts; this allows easy storage and fetching, but also allows for pollutants and pathogens to settle out of the water. In Accra the average per capita production of wastewater is approximately 50 litres per day, indicating that the flow of water from residents to urban vegetable plots informally contributes to the wastewater treatment of 225,000 residents (Figure 3). In principle, these 225,000 residents – some 14 per cent of the population – currently have a functioning wastewater treatment system that is not disposal oriented, but turns wastewater into an asset. This number is probably larger than the one served through sewerage.

The reduction of water pollution in Accra is critical for the environmental health of the city, as well as for its economic and nutritional well-being. Accra's largest wetland, the above-mentioned Korle Lagoon, which once supported a thriving fishery, still receives over 60 per cent of Accra's untreated greywater and wastewater, in addition to solid waste. Subsequently, the lagoon's drainage capacity has been significantly

1 bed = 6m²

1 watering can = 15L
10 watering cans / day / bed = 150L

Approx.
1000 beds / ha

75ha irrigated
vegetable cultivation

11,250,000L
urban water / day

Accra's population of 1,600,000 generates 80,000,000L of wastewater a day
Accra's 75ha of vegetable plots absorb 11,250,000L of urban water a day

Figure 3 | Calculation of the contribution of urban agriculture to wastewater reuse in Accra

compromised, leading to frequent flooding of low-lying areas with raw sewage, in addition to creating prime breeding grounds for disease-causing agents, especially mosquitoes. The lagoon has for many years been listed as one of the most polluted places globally (International Development Research Center, 1996), and its ongoing rehabilitation has required significant efforts and investment (Boadi and Kuitunen, 2002). An increased treatment of wastewater through urban agriculture would reduce the environmental burden and also contribute to the protection of local fisheries.

Solid waste management

Along the River Onyasia that flows in Accra east–west through Dzorwulu into the neighbourhood of Alajo, urban vegetable plots line the banks, tapering off to the west of Achimota Road. Beyond this point, buildings edge dangerously close to the steeply sloped and garbage-lined riverbank. Refuse dumping rarely occurs on urban agriculture sites; residents do not dump on land that is being consistently cultivated and the farmers themselves clear any debris that is deposited. On land that would otherwise be left vacant, urban agriculture prevents solid waste dumping and illegal development as farmers serve as informal caretakers for landowners. The ability of urban agriculture to prevent refuse dumping in and along major water

bodies clearly impacts on municipal planning goals; as solid waste clogging the system decreases, the volume of water that the drainage network can contain increases. The capacity of Accra's drainage system to manage stormwater during major rain events has been critically diminished by the collection of refuse in drains, clogging even major stream, as well as by illegal building in the flood plain.

Urban agriculture along streams and rivers in Accra is already serving as a deterrent for illegal dumping and development, minimizing the high risks and costs associated with clearing drains, removing illegal structures, and seasonal floods. By strategically encouraging urban agriculture along river and stream banks that are not suitable for development, the city of Accra could proactively limit the amount of dumping and development that occurs along these waterways. This approach needs to be coupled with increased access to formal waste disposal options so that solid waste dumping is not merely relocated.

While urban farms already discourage refuse dumping, their contribution to urban solid waste management could expand. Urban agriculture sites could serve as community destinations for organic solid waste collection and composting in order to locally manage solid waste and improve soil fertility on farms. Successful examples of this community-based solid waste management strategy include the projects of Waste Concern in Bangladesh (www.wasteconcern.org). The decentralized and heterogeneous nature of urban agriculture sites in Accra (i.e. physical character, tenure security, adjacent land use) favour decentralized community-based approaches, which can be expanded to other service potentials of urban agriculture (Figure 4).

Disaster management

Although flooding has always occurred in Accra, urbanization and highly probably also climate change have increased the occurrence and severity of floods, creating health and economic risks for urban communities (Douglas *et al.*, 2008). Dumping refuse and building in waterways augments the magnitude of floods by blocking water flow (Sam, 2002; Duodu, 2007). As noted above, cultivated sites discourage dumping and illegal encroachment, while farmers function as caretakers of the land. By limiting dumping and development along waterways, urban agriculture decreases human health and economic costs associated with urban floods.

In addition to expanding drainage networks in Accra, a common planning strategy involves channelizing and concretizing rivers and streams. While this intervention may solve a local flooding problem, it may also exacerbate urban flooding, particularly downstream (Parker, 2007). Water cannot percolate through the concrete into the riverbed; the absence of vegetation also decreases percolation while it increases the volume and speed of floodwaters downstream. A current scheme to channelize 1.8km of the Onyasia River, north of its Accra's Odaw River and south of the

Dzorwulu urban agriculture site, illustrates the high cost of mismanaging Accra's waterways, as well as the potential dangers of channelizing this river. As part of Ghana's 'Second Urban Environmental Sanitation Project', this channelization will require the removal of houses built within a demarcated floodway and the concretization of the river (Government of Ghana, 2003).

In coordination with the very expensive and disruptive strategy of retroactively managing urban development along waterways, a proactive approach is needed to support the promotion of buffer strips reserved for urban agriculture and forestry (Smit *et al.*, 1996) (Figure 5). For example, Ghana's Water Resources Commission initiated studies under the second phase of the Water Supply and Sanitation Programme (WSSP-II), 2004–2008, to develop a uniform buffer policy for all riverbanks, reservoirs and lakes to protect water resources and discourage inappropriate development. The establishment of urban agriculture within these buffer zones could contribute to the enforcement of these new regulations, though it has so far been considered as a threat, e.g. due to the use of fertilizers. Further study is required here to verify the contribution of the normally used manure compared to what already enters the streams from domestic sources. However, it is likely that a buffer zone of vegetable beds will prevent significantly more pollution than it might contribute.

Health care

Urban agriculture also contributes in other ways to urban health, in particular through food supply. Both open-space and backyard farming increase the availability of staple crops, such as cassava, yams and maize within Accra, while open-space farming increases access to nutrients from vegetables (IDRC, 2001). Every day in Accra, 200,000 people consume urban-produced vegetables at fast food kiosks and canteens – this indicates a daily impact of urban agriculture on 12.5 per cent of the population (Obuobie *et al.*, 2006). In addition to vegetables, open-space farming in Accra also produces herbs that are used in traditional medicines for diseases such as malaria; compared to food supply from rural areas, however, urban farming contributes only larger shares of very selective commodities, such as perishable vegetables (Drechsel *et al.*, 2007).

As semi-public spaces, urban agriculture sites in Accra are often adjacent to soccer pitches and include networks of pedestrian circulation. Both soccer

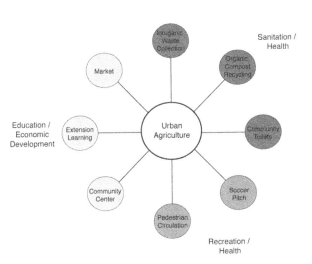

Figure 4 | Opportunities for expanding services and community benefits of urban agriculture in Accra

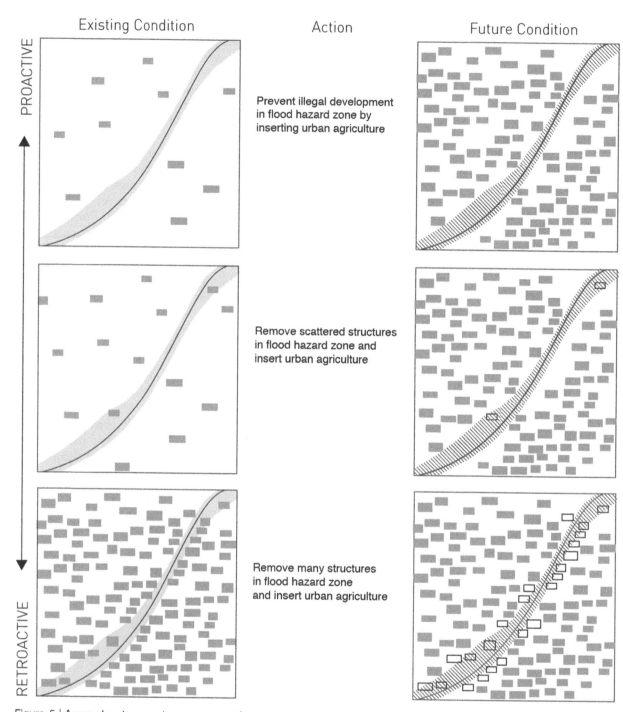

Existing Condition — Action — Future Condition

PROACTIVE → RETROACTIVE

Prevent illegal development in flood hazard zone by inserting urban agriculture

Remove scattered structures in flood hazard zone and insert urban agriculture

Remove many structures in flood hazard zone and insert urban agriculture

Figure 5 | Accra devotes great resources each year to retroactively removing illegal structures from flood hazard zones. Proactively siting urban agriculture in these zones could be an innovative way to provide the multiple benefits of urban agriculture while also achieving urban planning goals

pitches and pedestrian circulation enhance opportunities for local exercise; furthermore, urban agriculture may also pose benefits for mental health. For both the farmer and local community, Accra's urban agriculture sites are predominantly separated from vehicular traffic, providing a rare relief from urban stress.

Poor percolation of excess waters into compacted urban soils promotes breeding grounds for mosquitoes.

While several studies in Ghana suggest that urban agriculture does not increase the occurrence of the malaria vector *Anopheles*, this potentially negative health impact of urban agriculture remains an important area of research (Afrane *et al.*, 2004; Klinkenberg *et al.*, 2005; Cole *et al.*, 2006).

Conclusion: urban agriculture as a service provider

While classic arguments used to promote urban agriculture, such as its contribution to food supply or food security, might only appeal to local government in times of civil war or financial crisis, there are opportunities to highlight and augment the contribution of urban agriculture to the daily crisis of urban sanitation.

Assessing the multiple sanitation and health benefits associated with urban agriculture presents significant arguments for critically evaluating and encouraging urban agriculture as buffer zones against the environmental pollution of surface water bodies while contributing to flood control and decentralized wastewater treatment. Although urban agriculture should not be promoted as a panacea to address various problems, it can well contribute to an integrated concept of diverse measures to enhance waste management and environmental protection and probably pay for itself through its agricultural productivity and reduced costs for the cleaning of gutters and streams.

Urban agriculture can thus contribute to any decentralized model of waste and stormwater management, and a more nuanced approach to strategically treating and reusing wastewater. Linking urban agriculture to municipal services such as wastewater treatment also increases the probability that urban farms, with all of their quantitative and qualitative benefits for cities, will be sustained in a rapidly developing city such as Accra. What is now needed are urban sanitation strategies that recognize the role of urban agriculture and support its possible leverage and contribution, e.g. to public health, particularly in those developing countries where conventional sanitation remains a grave challenge.

Note

1. This article mistakenly implies that the James Town UASB sewage treatment plant has functioned continuously since construction in 2000.

References

Afrane, A., Klinkenberg, E., Drechsel, P., Owusu-Daaku, K., Garms, R., Kruppa, T., 2004, 'Does irrigated urban agriculture influence the transmission of malaria in the city of Kumasi, Ghana?', *Acta Tropica* 89, 125–134.

AMA (Accra Metropolitan Authority), 2006, *Revenue Mobilization, Sanitation and Waste Management Strategy 2006–2009*, Government of Ghana, Accra, Ghana.

Awuah, E., Abrokwa, K. A., 2008, 'Performance evaluation of the UASB Sewage Treatment Plant at James Town (Mudor), Accra', *Proceedings of the 33rd WEDC International Conference, Accra, Ghana*, 20–25.

Boadi, K. O., Kuitunen, M., 2002, 'Urban waste pollution in the Korle Lagoon, Accra, Ghana', *The Environmentalist* 22, 301–309.

Cofie, O., Abraham, E. M., Olaleye, A. O., Larbi, T., 2008, 'Recycling human excreta for urban and peri-urban agriculture in Ghana', in: L. Parrot, A. Njoya, L. Temple, F. Assogba-Komlan, R. Kahane, M. Ba Diao, M. Havard (eds), *Agricultures et développement urbain en Afrique subsaharienne. Environnement et enjeux sanitaires*, L'Harmattan, Paris, 191–200.

Cole, D. C., Bassil, K., Jones-Otazo, H., Diamond, M., 2006, 'Health risks and benefits associated with UA: impact assessment, risk mitigation and healthy public policy', in: A. Boischio, A. Clegg, D. Mwagiore (eds), *Health Risks and Benefits of Urban and Peri-Urban Agriculture and Livestock (UA) in Sub-Saharan Africa: Resource Papers and Workshop Proceedings*, Urban Poverty and Environment Series Report #1, IDRC, Ottawa, Canada.

Douglas, I., Alam, K., Maghenda, M., Mcdonnell, Y., Mclean, L., Campbell, J., 2008, 'Unjust waters: climate change, flooding and the urban poor in Africa', *Environment and Urbanization* 20, 187–206.

Drechsel, P., Dongus, S., 2009, 'Dynamics and sustainability of urban agriculture; examples from sub-Saharan Africa', *Sustainability Science*, doi:10.1007/s11625-009-0097-x.

Drechsel, P., Graefe, S., Sonou, M., Cofie, O., 2006, *Informal Irrigation in Urban West Africa: An Overview*, Research Report 102. IWMI, Sri Lanka.

Drechsel, P., Graefe, S., Fink, M., 2007, *Rural–urban Food, Nutrient and Virtual Water Flows in Selected West African Cities*, International Water Management Institute, Colombo, Sri Lanka, 35p (IWMI Research Report 115) [available at www.iwmi.cgiar.org/Publications/IWMI_Research_Reports/PDF/pub115/RR115.pdf].

Drechsel, P., Cofie, O., van Veenhuizen, R., Larbi, T., 2008, 'Linking research, capacity building, and policy dialogue in support of informal irrigation in West Africa', *Irrigation and Drainage* 57, 1–11.

Drechsel, P., Scott, C. A., Raschid-Sally, L., Redwood, M., Bahri, A., 2010, *Wastewater Irrigation and Health: Assessing and Mitigation Risks in Low-Income Countries*, Earthscan-IDRC-IWMI, London, p. 400.

Duodu, C., 2007, 'Water, water, everywhere and not a drop in the dam!', *New African* 464, 54–55.

Flynn-Dapaah, K., 2002, *Land Negotiations and Tenure Relationships: Accessing Land for Urban and Peri-Urban Agriculture in Sub-Saharan Africa*, IDRC Cities Feeding People Series, Report 36, IDRC, Ottawa, Canada.

Ghana Ministry of Health, 2007, *Ministry of Health: Facts and Figures* [available at www.moh-ghana.org/moh/docs/

health_service/SUMMARYOFTOPTWENTYCAUSESOF
OUTPATIENTMORBIDITY2007.pdf].

Ghana Statistical Service, 2004, *Ghana Demographic and
Health Survey 2003*, GSS, Noguchi Institute, ORC Macro,
Calverton, MA.

Government of Ghana, 2003, *Second Urban Environmental
Sanitation Project (UESP II): Environmental and Social
Assessment*, Government of Ghana, Ministry of Local
Government and Rural Development, Accra, Ghana.

Government of Ghana, 2008, *Preliminary National Environmental
Sanitation Strategy and Action Plan (NESSAP)*, MLGRDE,
Government of Ghana, Accra, Ghana.

IDRC (International Development Research Center), 1996,
Ghana: The Nightmare Lagoons [available at http://archive.idrc.
ca/books/reports/e234-13.html].

IDRC (International Development Research Center), 2001, 'Urban
agriculture: food security and nutritional status in Greater Accra
(Accra, Ghana)', in: *Improving Nutrition Through Home
Gardening: A Training Package for Preparing Field Workers in
Africa*, Nutrition Programmes Service, Food and Nutrition
Division, FAO (UN), Rome.

IWMI (International Water Management Institute), 2009,
*Wastewater Irrigation and Public Health: From Research to
Impact – A Road Map for Ghana*, A report for Google.org.
IWMI, Accra, Ghana.

Klinkenberg, E., McCall, P., Hastings, I., Wilson, M., Amerasinghe,
F., Donnelly, M., 2005, 'Malaria and irrigated crops, Accra,
Ghana', *Emerging Infectious Diseases* 11, 1290–1293.

Nhapi, I., Gijzen, H. J., 2004, 'Wastewater management in
Zimbabwe in the context of sustainability', *Water Policy* 6,
501–517.

Obuobie, E., Keraita, B., Danso, G., Amoah, P., Cofie, O.,
Rschid-Sally, L., Drechsel, P., 2006, *Irrigated Urban Vegetable
Production in Ghana, Characteristics, Benefits and Risks*, IWMI/
RUAF/CPWF, Accra, Ghana.

Parker, R., 2007, *IEG World Bank: Independent Evaluation
Group. Development Actions and the Rising Incidence of
Disasters*, Evaluation Brief 4, June, World Bank, Washington,
DC.

Prüss-Üstün, A., Bos, R., Gore, F., Bartram, J., 2008, *Safer Water,
Better Health: Costs, Benefits and Sustainability of
Interventions to Protect and Promote Health*, World Health
Organization, Geneva.

Sam Jr., P. A., 2002, 'Are the municipal solid waste management
practices causing flooding during the rainy season in Accra,
Ghana, West Africa?', *African Journal of Environmental
Assessment and Management* 4, 56–62.

Satterthwaite, D., McGranahan, G., 2007, 'Providing clean water
and sanitation', in: *State of the World 2007: Our Urban Future*,
Worldwatch Institute, Washington, DC.

Smit, J., Ratta, A., Nasr, J., 1996, *Urban Agriculture: Food, Jobs,
and Sustainable Cities*, United Nations Development
Programme, New York.

Songsore, J., Nabila, J.S., Yangyouru, Y., Amuah, E.,
Bosque-Hamilton, E. K., Etsibah, K. K., Gustafsson, J., Gunnar,
J., 2005, *State of Environmental Health. Report of the Greater
Accra Metropolitan Area 2001*, Ghana Universities Press,
Accra, Ghana.

World Health Organization, 2006, 'Guidelines for the safe
use of wastewater, excreta and grey water', in:
Wastewater Use in Agriculture (Vol. 2), World Health
Organization, Geneva.

Agriculture on the edge: strategies to abate urban encroachment onto agricultural lands by promoting viable human-scale agriculture as an integral element of urbanization

Patrick M. Condon[1], Kent Mullinix[2]*, Arthur Fallick[2] and Mike Harcourt[3]

[1] School of Architecture and Landscape Architecture, University of British Columbia, 394-2357 Main Mall, Vancouver, BC V6T 1Z4
[2] Institute for Sustainable Horticulture, Kwantlen Polytechnic University, 12666-72nd Avenue, Surrey, BC V3M 2MB
[3] Centre for Sustainability-Continuing Studies, University of British Columbia, 800 Robson Street, Vancouver, BC V6Z 3B7

In the Greater Vancouver region (Canada) tensions exist where urbanization encroaches onto agricultural land. A recently issued white paper proffered ideas to stimulate discussion on land-use plans and public policies to encourage and enhance agriculture while accommodating a doubling of the region's population. It evoked a visceral response from local and regional politicians, planners and agrologists who saw it as an heretical attempt to undermine land conservation. Proponents saw innovative strategies to ameliorate entrenched antipathy between competing perspectives. The core arguments and corresponding critique, outlined in this paper, bring to light elements of a broader debate about the vitality and sustainability of agriculture in British Columbia, as elsewhere, centring on issues of food security (supply) and food sovereignty (control) within two competing agricultural paradigms: human-scale agri-food systems and conventional industrial agri-business. Municipal enabled agriculture (MEA) is advanced as a catalyst for the full integration of the agri-food system within the planning, design, function, economy and community of cities and vice versa. MEA can make significant contributions to local and regional economies and has the potential to alter the way communities are designed to reduce unsustainability, planned to incorporate resilience, and organized so that they flourish socially and culturally.

Keywords: food security, food sovereignty, human-scale agriculture, planning, sustainability, urban encroachment

Introduction

Questions of sustainability have come to dominate much of the recent discourse regarding the future of post-industrial society in general and the security of our agri-food systems in particular. Concomitantly, and for the first time in history, the majority of the world's population is urbanized (United Nations, 2007). In Canada, for example, only 3 per cent of the population resides on farms and only 1.4 per cent of the population is engaged in farming (Agriculture and Agrifood Canada, 2002). In other words, the vast majority of Canadians have little or no meaningful connection to their agri-food system – a consequence of 20th-century industrialization and economic globalization.

While there is a growing recognition of the limitations and challenges that this path is having economically, socially and ecologically, many remain fairly ignorant of the ecological principal and ecological processes that affect every aspect of their daily life. People are becoming increasingly sequestered in cities and insulated from ecological engagement and awareness. More importantly, too many are generally unaware of the ecological burden being imposed upon the earth's resources and systems, despite the fact that most

*Corresponding author. Email: kent.mullinix@kwantlen.ca

INTERNATIONAL JOURNAL OF AGRICULTURAL SUSTAINABILITY 8 (1&2) 2010

PAGES 104–115, doi:10.3763/ijas.2009.0465 © 2010 Earthscan. ISSN: 1473-5903 (print), 1747-762X (online). www.earthscan.co.uk/journals/ijas

would agree that human activity profoundly influences the local, regional and global ecological functions which human welfare is dependent upon. There is perhaps an encouraging trend. We are seeing an expanding discussion of sustainability issues that examines the strategic significance of food security (supply) and food sovereignty (control) (City of Richmond, 2003; Kent Agriculture Advisory Committee, 2004; American Planning Association, 2007; District of Maple Ridge, 2009).

In southwest British Columbia (BC), Canada, as elsewhere, there is a growing awareness that the combined effects of peak oil, peak water, climate change, rapid urbanization, continued population growth as well as the current status, configuration and dominance of conventional industrial agriculture have the potential to undermine the resilience of our cities, threaten food security and ultimately result in an agri-food system that is not sustainable (Rosenweig *et al.*, 2000; Kimbrell, 2002; Heffernan, 2005; British Columbia Ministry of Agriculture and Lands, 2006; Campbell, 2006; Heinberg, 2006; Barlow, 2007; McKibben, 2007; Patal, 2007; Garnett, 2008; Roberts, 2008). Evidence of these forces converging was felt in 2008 with an inflation rate of 1.2 per cent overall, while food costs rose 7.3 per cent, cereal products 12.4 per cent and fruits and vegetables a staggering 26.9 per cent (CBC News, 2008).

Efforts to promote a sustainable agri-food system through the expansion of urban and peri-urban agriculture in our region range in scale from grass-roots activism such as community gardens, SPIN farming (small plot intensive farming) and farmers markets, through design parameters such as green roofs and edible landscaping, to public policy initiatives such as the City of Vancouver's Food Policy Council, (Mendes, 2006), Sustainability Charters proclaimed by several municipalities, Metro Vancouver's Regional Growth Strategy (Metro Vancouver, 2009) and the Agricultural Land Reserve (ALR) legislation enacted by the Government of British Columbia (Provincial Agriculture Land Commission, 2002). In fact, the BC Ministry of Agriculture and Lands' recent publication: *British Columbia Agriculture Plan: Growing a Healthy Future for B.C. Families* (British Columbia Ministry of Agriculture and Lands, 2008a) calls for enhanced community-based/local food systems, addressing food security through diverse local production, environmental stewardship/climate change mitigation and bridging the urban–agriculture divide.

Results are mixed thus far and collectively have yet to yield a resilient, adaptive and sustainable agri-food

system or to cause a consistent, coherent or comprehensive strategy to emerge. We believe the answer lies in part in envisaging and building a municipal-focused agriculture sector in which agriculture and urbanity are inextricably linked via planning and economic strategy (Esseks *et al.*, 2008). We contend that human-scale municipal-focused agriculture should form the basis of a bio-regional agri-food system as a necessary precondition for creating local and regional food security (supply) and food sovereignty (control). This perspective raises several key research questions for us, relating to how food systems should be configured to contribute to more sustainable, liveable urban centres, while enhancing the agri-food sector:

- How might urban and peri-urban agriculture be tied directly into the ecological and social function and economic vitality of our cities?
- How can human-scale agri-food production realize sustainability objectives while contributing to lessening the urban ecological footprint?
- How can human-scale, urban-linked agri-food systems contribute to the social fabric of our cities providing opportunity for productive, healthy human engagement and enterprise?

We explore these questions through a combination of a recent case study and field-based research which is being conducted with private sector, municipal government and community-based partners, in a geographic arena that is struggling to contend with significant growth management challenges.

Metro Vancouver and the Agricultural Land Reserve

In 2009 a white paper entitled *Agriculture on the Edge* (Condon and Mullinix, 2009) was written by two seasoned academics and introduced at a summit of invited regional leaders representing various sectors and interests by a former premier of the Province of British Columbia. The objective of the paper and summit was to stimulate discussion around the inherent tensions that currently exist between rapid urbanization and the encroachment onto land zoned for agriculture in the Metro Vancouver region, and to call attention to the urgent need to abate urban encroachment on agricultural lands by promoting viable agriculture as an integral element of urbanization.

The authors examined the dynamic interplay of competing forces (urban growth vs. preservation of farmland) 'at the edge' and proffered ideas that were intended to stimulate discussion on potential formulations of land-use plans and public policies that would encourage and enhance agriculture while simultaneously accommodating the anticipated doubling of the region's population over the next 20–30 years (Baxter, 1998). Their concept paper included a provocative suggestion that a 500m zone of land at the interface of the urban and agricultural lands could be considered for an innovative approach to creating enhanced agriculture. The proposal was to capture the 'value lift' on a 200m corridor of the land after it had been rezoned to allow urban development, and use the monies derived from this 'lift' to stimulate and finance enhanced agriculture through a form of 'community trust farming'.

The paper evoked a visceral response from local and regional politicians, planners and agriculturists as well as support from proponents who saw in the proposal innovative strategies to ameliorate the entrenched antipathy between competing perspectives on how growth should be managed in the region and how to integrate *agri-culture* into *urban-culture*. Opponents saw the paper as an heretical attack on scarce and precious farmland and their preservation strategy. The effective rallying cry of a vocal campaign to preserve the ALR has become: no buildings on farmland! While effective as a clarion call, it has done nothing to advance the dialogue.

The essence of *The Edge* paper and the corresponding critique are outlined in the next section. As a case study, it highlights elements of a broader dynamic debate about the vitality and sustainability of agriculture in BC, as elsewhere, that centres on issues of food security (supply) and food sovereignty (control) within two competing agricultural paradigms: human-scale agri-food production and conventional industrial-agricultural production.

We offer a contribution to this dynamic debate with exploratory ideas from an emergent research and development agenda at the Institute for Sustainable Horticulture (Mullinix *et al.*, 2008, 2009) which examines the potential for municipal enabled agriculture (MEA) as a catalyst for the full integration of the agri-food system within the planning, design and function of cities, and vice versa. MEA is defined as agricultural enterprise that is human scale, ecologically sound, in and around cities, for and by communities. We contend that MEA can make a significant contribution to local and regional economies (creating jobs, real wealth and the next generation of urban farmers), and has the potential to alter the way communities of the future are designed to reduce unsustainability (Ehrenfeld, 2009), planned to build-in resilience (Southlands

in Transition, 2009) and organized so that they flourish socially and culturally as sustainable communities.

Agriculture on the edge

The Metro Vancouver region is an amalgamation of 21 cities and municipal districts, encompassing 282 million ha, including 41,000ha of farmland (Figure 1), with a population of 2.1 million. The population is expected to double by 2040. Metro Vancouver has a long and rich agricultural heritage and remains an important element of BC's agriculture sector, currently generating 25 per cent of gross farm receipts from 14 per cent of the agricultural land base. Smaller, family owned and operated farms dominate (88 per cent are less than 26ha), but farm numbers have declined by 25 per cent in the last 10 years. The average age of the region's farmers is 55 years, fewer children are opting to carry on the family tradition, and farmland has become prohibitively expensive for those who are interested in becoming the next generation of farmers (Metro Vancouver, 2007; Pynn, 2008).

The ALR is a precedent-setting provincial regulation intended to conserve agriculture land and enhance agriculture in BC. For the last 30 years it has been a de facto urban growth boundary, resulting in our metropolitan areas being significantly more compact than most in North America. While this has been a positive outcome, ALR land values have risen to $100.000 (Cdn) or more per acre – a cost that cannot be serviced by typical farm receipts.

Although total provincial ALR lands have experienced no net loss (SmartGrowth BC, 2004), a significant portion of prime ALR designated agricultural land within the Metro Vancouver region has been swapped for lesser quality lands in distant regions of the province (Campbell, 2006; Cavendish-Palmer, 2008). Much of the Metro Vancouver region land originally designated to be within the ALR is fragmented and has been abandoned for agricultural purposes, we believe, either because it is too small or otherwise inappropriate for 'industrial' scale agriculture use, because it is being held for speculation, or because it is a land-endowed 'country residence'. Five interrelated factors contribute to this dynamic, each of which threatens the goal of preserving productive agriculture land and providing a significant degree of regional food security:

1. Development pressures are mounting as nearly all of the easily developed sites outside of the ALR are either 'built out' or planned for building.

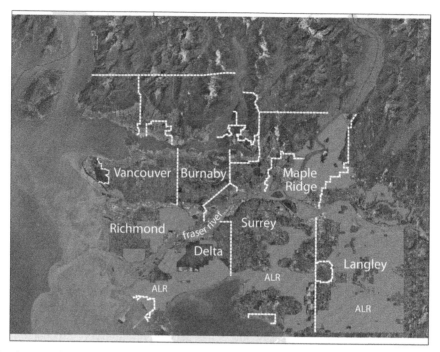

Figure 1 | Satellite photograph of the greater metropolitan Vancouver region of British Columbia, Canada
Major municipal boundaries are delineated by dashed white lines. Urban/suburban areas are evident as dark and mottled with building, roads, etc., while ALR lands are light grey. Note ALR lands are completely surrounded by urban/suburban development. The extensive and significant fragmentation within ALR lands by such things as roads, golf courses, industrial facilities and development is not evident. Mountains are to the north and the Fraser River flows west to the Pacific Ocean.
Source: Metro Vancouver, with permission.

Developers and local politicians feel there is limited potential for development within existing urban zones (e.g. they consider infill, intensification of existing neighbourhoods or the wholesale reconstruction of existing urban areas to be impractical). Without a substantial change in both the development community and the political culture, this proclivity will probably persist.

2. The preponderance of ALR lands near urbanized areas have been purchased at values or are valued at orders of magnitude higher than can be justified by any form of conventional agriculture utilization (reaching $100,000 per acre or more). ALR land is clearly being purchased expressly to hold for speculative investment purposes, fully expecting the ALR to break down in the near future. What is more, land speculators are afforded a tax incentive as agricultural lands are taxed on an advantageous scale with very easily satisfied agriculture production and income generation requirements. This makes the cost of holding these lands much more affordable for land development speculators (Penner, 2008).

3. Municipalities have been granted rights to review ALR exclusion requests before the Agricultural Land Commission (a provincially appointed adjudicating body) makes its decision. Local councils have proved more likely to allow exclusions than have Provincial boards. Local politicians feel pressure to release lands more acutely than distant Provincial regulators, and apparently find it more difficult to deny an application from someone they may know and who may have political influence. Many concerned citizens are now convinced that municipal exclusions drive the system.

4. The majority of Metro Vancouver agricultural lands that are farmed primarily produce crops (blueberry, cranberry, raspberry, vegetables) for volatile, low-margin commodity markets in the global agri-food system (Table 1). For these, increasingly marginal (and insufficient) returns on investment are often realized (Morton, 2008). Some land is used for producing supply-managed commodities (milk, eggs, poultry) as well as high-value crops (e.g. mushrooms and nursery plants). For the former, strict controls on competition and price structure largely mitigate

Table 1 | Crop production in Metro Vancouver, 2006

Crop type	Land area (ha)			% Change 1996–2006
	1996	2001	2006	
Berries	**3300**	**3940**	**4643**	**29%**
Blueberries	1506	1746	2734	45%
Cranberries	1218	1505	1503	19%
Strawberries	204	227	208	2%
Raspberries	318	200	198	−38%
Vegetables	**2639**	**3175**	**3025**	**13%**
Potatoes	2097	2085	2285	8%
Green/wax beans	444	627	804	45%
Sweet corn	366	469	405	10%
Squash/pumpkin/ zucchini	186	308	320	42%
Lettuces	271	282	213	−21%
Cabbage	161	197	97	−40%
Carrots	182	259	196	7%
Spinach	29	58	48	40%
Celery	46	20	7	−85%
Rutabaga/turnip	39	44	45	13%
Chinese cabbage	74	96	105	30%
Nursery crops	**1113**	**1235**	**1192**	**7%**

Source: Metro Vancouver (2007) Census Bulletin #2 Census of Agriculture.

has become a zone of planning contention and land-use conflict. Those who live on the urban side of an arbitrary boundary ('hard' edge) affected by regulation often consider the practices of industrial-scale agri-business as an impediment to their quality of life (Figure 2). At the same time they ascribe high value to the 'protected agricultural lands' for their aesthetic importance as pastoral open space. Meanwhile, those attempting to farm the lands on the agricultural side of the arbitrary boundary feel threatened by further urban encroachment that brings with it the pressures of speculation on land value and operational conflicts that often arise with industrial-scale farm practices.

Without doubt, designated ALR lands in southwest BC, particularly those in the Metro Vancouver region, are under substantial threat from urban expansion and other non-agricultural uses (SmartGrowth BC, 2004; Campbell, 2006; Cavendish-Palmer, 2008; Metro Vancouver, 2009). The strategy of relying exclusively on this regulatory tool to ensure land is available for food production and to provide a buffer between agricultural and urban lands has significant limitations, is politically polarizing, and fails to advance regional food security or food sovereignty. Incorporating urban design principles and sustainable land-use practices that integrate human-scale food production with nearby urban settlement (particularly at the urban–ALR 'edge') may be a more effective way to resolve this seemingly intransigent problem. Human-scale urban and peri-urban agriculture and related efforts to support the ALR are necessary to reduce unsustainability and contribute to resilience in

competitive market forces. Other ALR designated holdings are utilized for low intensity or pseudo-agriculture (e.g. horse stables and 'hobby farms') or produce low value per acre crops such as Christmas trees. The majority of Metro Vancouver ALR lands are not farmed for high value local/ regional markets which generally provide a high return on investment (British Columbia Ministry of Agriculture and Lands, 2008b). As a result, farms continue to consolidate, fewer and fewer individuals and families farm, and young persons eschew farming/agriculture (Metro Vancouver, 2007). These factors significantly undermine efforts to actualize regional food security and to maintain the economic vitality of the agriculture sector.

5. The interface between lands designated for agricultural use (either in the ALR or zoned for agriculture) and the adjoining developed lands at the urban edge

Figure 2 | Typical hard edge at the urban settlement and agriculture interface in Metro Vancouver, British Columbia, Canada; it is perfectly suited for the proposed planning tool and human-scale, municipal enabled agriculture

BC, and are prerequisites to creating sustainability and reconnecting urbanity to its roots in the land (Freyfogle, 2001; Carlson, 2008).

Singly and collectively, the aforementioned factors are counterproductive to the long-term preservation of regional agriculture lands and to the creation of a sustainable bio-regional agri-food system. The problem clearly requires rectifying. In doing so we cannot ignore the fact that most of the provincial ALR land at greatest risk today is in fast-growing metropolitan regions and under present circumstances is contributing very little to regional food security. They are lands solely protected by ALR regulations, but not regulated for their agricultural productivity or contribution to regional food security.

New planning tools needed

The intent of the ALR was to promote viable farming, not to function as an urban growth boundary (Smith, 2007). To be fair, this far-thinking and precedent-setting legislation has served to conserve agriculturally useful lands and keep Metro Vancouver more compact than most other North American metropolises. However, it has not facilitated a cessation of urban encroachment by any means and falls far short of promoting food security, food sovereignty and agriculture sector viability. A reasonable trajectory for the ALR is that it is likely to be further compromised, in fits and starts, a few acres at a time, until very little agriculture in Metro Vancouver remains. As such a more substantial and creative action to conserve farmland and promote a viable agri-food sector may be justified. Any such strategy should enhance urban settlement as well.

Such an action would somehow take all five challenges listed above and recombine them to create the economic and social opportunities necessary to effectively solve this conundrum. We elucidate one seemingly radical but practical solution below. We do not offer it as a panacea but as one potential strategy, valuable in (hopefully) an array of strategies and models.

A planning tool proposed

What follows is the outline of a six-element planning strategy to complement and strengthen the ALR. It includes a new zoning designation, transfer of some urban edge land and value lift to agriculture and the public sector, integration of human-scale agriculture with urban dwellers, promotion and support of a new

and critically important agriculture and economic sector and a tangible contribution to regional food security. It assumes that this can be accomplished without public sector tax dollars. The elements of this strategy are:

1. The province, the region, and its member municipalities establish a planning zone of up to 500m wide at the interface between urban and agricultural or preservation lands. Such lands are to be used for both urban and agricultural purposes. Urban uses would presumably be held to 100–200m, with the remainder of the planning zone restricted under covenant for intensive agriculture aimed at local markets.

2. This new band could be rezoned for medium- to high-density living on developed portions. For the sake of this discussion we shall assume a yield of 60 dwelling units per net acre, allowing for significant return on developer investment. Sixty dwelling units per net acre ('net' meaning the number of units per acre on just the development parcels) or 40 dwelling units per acre gross ('gross' meaning the number of units per acre when roads are included in the calculus) would exceed 10 dwelling units per 'double gross' acre ('double gross' meaning the average density when open spaces and agricultural lands are also included in the calculus). Ten to 15 dwelling units per double gross acre is usually considered the minimum density necessary to support viable transit services and local commercial services.

3. Protect, legally and in perpetuity (e.g. via covenant and/or land trust consignment), two-thirds of this land (relinquished by the owner/developer) exclusively for agriculture (Pringle, 1994; Gillon et al., 2006). It may be desirable that designated agriculture land ultimately comes under the ownership of the associated municipality. If it does we refer to this arrangement as Community Trust Farming.

4. Lease (very favourably) these agricultural lands to agricultural entrepreneurs and stipulate they be farmed exclusively for local/regional markets, thus contributing to the sustainability of our communities and to genuine regional food security. Require that labour-intensive, high-value crops and value-added products (e.g. organic, direct marketed) be produced and that labour-intensive highly productive and sustainable production practices be utilized as opposed to capital and input (pesticides, fertilizers, mechanization) intensive industrial methods.

5. Relegate the oversight of these lands to a non-governmental organization (NGO), community/resident associations, or professional consulting agrologists under deed restrictions that would compel use as stated above. It may also be that municipalities will hire professional agriculturists to manage and/or farm these lands.

6. Endow these lands with funds garnered at the time of land sale to support local and sustainable agriculture in perpetuity. Through provincial authorization, local governments already exact a 'Development Cost Charge' from development projects, as a means to finance associated public infrastructure and services requirements associated with municipal growth. Per this scheme the local/regional agri-food system becomes an integral element of municipal growth. Thus it seems reasonable that Development Cost Charge structures could be modified and appropriately used to support the creation and stewardship of municipally focused agri-food system components.

The key

The economic basis for this concept is simple. When lands shift from agricultural to urban use the land values increase substantially. The 'lift' in value can be huge, from a $40,000 per acre value as agricultural land to over $1 million per acre as urban land (depending on location and specific development capacity). Typically, the public act of allowing this to occur generates a huge shift in value to land owners and land speculators only, while ignoring or subverting many strongly held citizenry interests, including food security, curtailing urban sprawl, preventing agricultural land loss, and having a viable agriculture sector.

The public sector, however, has the right and ability to change this dynamic by capturing a large portion of the value lift at the time of rezoning application, and using it toward desired ends, which might be Community Trust Farming (Condon and Mullinix, 2009) or another mechanism supporting local food security. If half of the aforementioned value shift is captured through development fees it would generate, for the sake of our discussion, up to or perhaps over $500,000 per acre. Using this figure, each 10-acre parcel would then provide $5 million to endow the activities of local agriculturists and Community Trust Farming land management. Invested value capture would generate roughly $200,000 per annum (depending on contemporary interest rates) to support each 5–7 acres of labour-intensive

and nutrition-rich agriculture operation held and operated in trust. It may also be that some of this captured value can be used to support regional agri-food systems infrastructure and support such as farmers' markets, incubator kitchens and extension research and education support services. Given that intensive, ecologically sound, locally/regionally focused agriculture has difficulty competing economically in the current economic and global agricultural context, some level of support would be beneficial. However the payoff could be large (Kaufman and Bailkey, 2000; Goldberg, 2006).

Food products generated would, per stipulation, only be sold in local/regional markets, making healthy, wholesome, diverse and affordable foods available to a larger number of citizens and putting into place an infrastructure requisite for local/regional food security. The increased nutritional content of sustainably cultivated food crops, a hidden and far-reaching economic benefit to consumers and taxpayers, is now documented. For example organic fruits, vegetables, grains and meats have been routinely found to contain significantly higher levels of various vitamins, minerals and antioxidants (Davis *et al.*, 2004; Benbrook and Greene, 2008; Benbrook *et al.*, 2008).

A substantial and economically robust local/regional agri-food sector would result – one that supports entrepreneurship, small business, creates green jobs and contributes to the regional economy (Tunnicliffe, 2007; British Columbia Ministry of Agriculture and Lands, 2008c; Farmers Markets Canada, 2009; Illinois Local and Organic Task Force, 2009). These potential economic and social benefits cannot be overstated (Meter and Rosales, 2001; Korton, 2009). In addition to the straightforward benefits of regional economic diversification, most revenue generated from these farms would stay and circulate within the regional economy, multiplying in value and economic effect, rather than quickly leaving to distant corporate headquarters as is increasingly the case (Heffernan, 2005). Concomitantly, the nature of a community's agriculture sector profoundly influences its social and economic character. Communities dominated by smaller, family-owned farms and agriculturally related business, compared to ones dominated by consolidated, transnational agribusiness, have been found to have overall higher standards of living, lower crime and poverty rates, more retail trade and independent businesses and more parks, schools, churches, newspapers and citizen involvement in democratic processes (Goldschmidt, 1978).

Research also indicates increasing consumer support for small-scale regional/local farming, sustainably produced food products and a willingness to preferentially patronize them while paying a premium. In the United States, farms of less than 50 acres (human scale, direct market) and those over 2000 acres (consolidated agri-business) are the only ones prospering and increasing in number. All others are in decline (Kirshenmann, 2004; Kirshenmann *et al.*, 2004). In North America organic food is the only product category in retail food sales experiencing growth, and in Canada farmers market sales now exceed $3 billion annually (Archibald, 1999; Statistics Canada, 2008; Farmers Markets Canada, 2009). Consumers are now prepared economically and politically to support an agri-food system that is environmentally sound, promotes a sustainable and secure regional food system and contributes to building economically vital and socially coherent communities (Thompson, 2000; Ipsos Reid Public Affairs, 2008). In our proposed scheme, such an agri-food system would emerge without direct taxpayer support. Rather, the support would come exclusively from public capture of a portion of the value lift associated with rezoning and urban development.

Furthermore, the pattern of development could be configured such that the acreages closest to homes would be farmed in the most unobtrusive ways (i.e. labour intensive and reduced chemical use/noise) to reduce potential conflicts between residential uses and agriculture practice. As you move away from homes, larger scale and more mechanized, conventional agriculture would be more suitable. Thus a range of and appropriate complement of agriculture enterprise types could be accommodated in a regional agri-food system. In this new agriculture sector conventional farmers may find opportunity for economically advantageous diversification. Finally, even though new buildings might consume 25–40 per cent of a site that may have been previously allocated to farming (in reality now mostly fallow or leased for conventional farming of low yield/margin products), by requiring small-scale labour-intensive farming on the remaining acres it is likely that the agricultural productivity of these lands, in terms of caloric output and nutritional value, will be many times greater than before (McKibben, 2007).

Reaction to *The Edge* paper

In response to the concepts brought forth in the white paper and summit, the Agriculture Committee of Metro Vancouver directed staff to prepare an analysis for their consideration (Rowen and Duynstee, 2009). Metro Vancouver is an inter-municipal governing body of the Greater Vancouver Regional District,

charged with certain aspects of governance for the metropolitan area. In the report was acknowledgement of the proposed planning tool and paper/summit objectives. However, the analysis contended that it had already been determined that all anticipated growth (to '2041 and beyond') could be accommodated without any encroachment onto agriculture, green space or parklands, thus, seemingly, dismissing any need for such strategies or further discussion altogether. The larger objectives of integrating human-scale agriculture with urbanization and creating a soft interface eluded analysis. The central proposition as to how the edge planning concept might mitigate the intransigence and polarization around the debate while accommodating population growth and enhancing agriculture was dismissed.

The analysts took exception to various assessments of the nature and status of Metro Vancouver agriculture and ALR land utilization while acknowledging that 33 per cent of Metro Vancouver ALR lands are not used for agriculture. They contended (per 'anecdotal evidence') that all farmable lands were satisfactorily and fully utilized. The analysis countered that the nature of our agri-food sector and ALR land utilization patterns were appropriately directed by competitive free market forces and took exception to the notion that prescriptive (planning and policy) approaches to create a sustainable, regional agri-food system were appropriate. Further, the analysts were dubious of claims that intensive, human-scale agriculture could be more productive and valuable.

The core concept was found to be 'inconsistent' with 'sustainability principles' delineated in Metro Vancouver's regional growth strategy (Metro Vancouver, 2009) and supportive of only the first (of six) priorities put forth in Metro Vancouver's *Economic Strategy for Agriculture in the Lower Mainland* (Artemis Agri-Strategy Group, 2002), that priority being the protection of farmland. In the final analysis the planning concept and its objectives were deemed to work against broad regional growth strategy objectives.

The authors of *The Edge* paper were given the opportunity to respond in person to the in-house critique by the Agriculture Committee of Metro Vancouver, with a subsequent invitation to work with an ad hoc planning group to bring forward a revitalized Agriculture Plan for the region. This invitation affords the researchers the opportunity to press the case that new measures are called for if the ALR (with supporting public policies and land-use plans) is to ensure a sustainable bio-regional agri-food system with characteristics such as those delineated in our emergent concept of MEA.

Municipal enabled agriculture

Municipalities have a pivotal role to play in laying the foundations for a sustainable 21st-century urban-centred society in which human-scale agri-food systems are central. Currently, food has become little more than an urban sector throughput – it comes in (in untold quantities and forms) and its waste products (which are many) go out. We have little or nothing to do with its production, processing or marketing. We have no substantive relationship with this omnipresent, universal and fundamentally important aspect of our existence. Yet we know that the negative ecological and social implications of this system are many and great (Kimbrell, 2002; National Farmers Union, 2005a; Mullinix, 2003). It is our fundamental belief that municipalities hold the key to creating local/regional food systems because they represent the level of government that is best situated to effect the needed change, being closest to those for whom such a food system is intended (Mullinix *et al.*, 2008).

We use the concept MEA to describe the full integration of agriculture and the food system within the planning, development and function of our rapidly urbanizing communities. It is an agri-food system element intended to connect urbanites, in real and meaningful ways, to their environment and to a human enterprise that is undeniably crucial to their future well-being. It is a way of reducing vulnerability and dependence on an ecologically unsound and increasingly vulnerable agri-food system while simultaneously reducing our ecological footprint (Mullinix *et al.*, 2009; Sustainable Development Commission, 2009). It also has significant direct economic potential for BC by inverting the local–global dependency ratio. Based on our research and analysis, we contend that human-scale agri-food production can be an effective long-term strategy for strengthening and sustaining our local and regional economies, and enhancing agriculture and urban settlement. Further, we suggest that MEA represents a structured approach that can respond substantively to the economic and resource challenges that will increasingly beset BC particularly in regard to food security (defined in terms of supply), and food sovereignty (defined in terms of control) (Quayle, 1998).

Given western agriculture's record of resource dependence and depletion, ecological devastation and agricultural community devolvement, as well as the interface planning nightmares that beset our towns and cities, we believe that it is critical and timely to challenge the prevailing mindset that sees increased agricultural globalization, industrialization and separation

from urban settlement as a viable path to continue down (Hove, 2004). We depart from convention and suggest that human-scale, agri-food production represents an undervalued economic and community-building force that can transform how we design, plan and support our local communities. To this end, we are working closely with a number of progressive municipalities throughout BC to explore how MEA, as part of a bio-regional agri-food system strategy, can mitigate against the worst impacts of agriculture's environmental, economic and social challenges while at the same time demonstrating practical ways through which to build the workforce (the next generation of urban farmers), the work (food security and agri-food production) and the productivity (urban agriculture as a significant municipal economic engine).

Envisioning a preferred future

The industrialization and globalization of agriculture and the segregation of the vast majority from a relationship to their food production did not just happen by default. It has been planned and envisaged in boardrooms, design studios and through media manipulation (National Farmers Union, 2005b; Patal, 2007). In the same vein, the characteristics of a preferred agri-food system that can guide dialogue and inform regional planning innovation and implementation strategies must also be delineated. Our goal is to provide a focused, compelling and constructive position that will bring stakeholders together in common purpose, objective and effort in the hope that it might lead to a common vision around human-scale agri-food production as an integral part of resilient cities throughout Metro Vancouver and BC. The following describes elements of a preferred human-scale agri-food system within a sustainable bio-regional context:

1. Our agri-food system will be economically robust and will contribute significantly and directly to our local and regional economies.
2. Our urban-focused agriculturists will capture significantly more of the marketplace value of foods and products, at least to levels which afford reasonable rates of return.
3. Our agri-food sector will put many people to work in satisfying jobs. New jobs will be one measure of its economic and social viability and ultimate success.
4. Our agriculture will appeal to a new generation and represent a social and economic sector in which

they feel they can pursue rewarding, satisfying careers, live happy and meaningful lives and contribute to society in valued, personally rewarding ways.

5. Our agriculture engages our urban populace; it is not segregated from the vast majority. Rather it is a fully integrated and positive part of people's everyday lives; it connects people with the means to their sustenance, to the natural world and to each other. It fosters community.

6. Our urban and peri-urban agri-food system is environmentally sound, enhances our natural environment and contributes to the mitigation of environmental degradation. Farmers are recognized as skilled stewards of precious natural resources and farming as a critical, knowledge-intensive and noble profession.

7. Our agriculture will make healthy fresh foods readily available to all and contribute to the mitigation of diet-related disease.

8. Our agriculture, by virtue of how we support it, plan for it, integrate it with other aspects of life and urbanity and relate to it, will in and of itself be an impediment to land speculation, unbridled urban sprawl and loss of arable land. It will enhance urban environs and living.

9. Our region's urban-focused agri-food system will be diverse, multi-dimensional and strive to create and support many new models. Adaptability and resiliency lie in the diversity that affords a multitude of opportunities for response and adaptation.

10. Our agri-food system will genuinely address food security issues, ultimately focus on achieving regional food sovereignty, and thus contribute directly and in substantive ways to urban sustainability.

One of the most significant challenges facing our research team is to demonstrate the credibility of our concept of MEA as well as our vision for a preferred human-scale agri-food system. We are currently working closely with a number of progressive municipalities to explore ways to implement MEA in practical ways. For example, we are laying the foundations for a series of municipally supported farm schools that will help build the next generation of farmers, create jobs and demonstrate how urban agriculture can be a significant economic driver for municipalities. We are also discussing ways by which these concepts can be incorporated into municipal and regional agriculture plans and implementation strategies. In particular, our discussions with senior municipal leaders are directed towards

identifying specific on-the-ground projects that will demonstrate various facets of MEA (e.g. community trust farming, incubator and community farms, community-supported agriculture enterprises, etc.) through a series of 'living laboratories'. We envisage the cumulative results of these initiatives being brought together within networked centres of excellence, culminating in a BC Centre for Human-scale Agriculture. Our current capstone project brings together both the conceptual ideas presented here and a detailed design schema in a substantial land holding (525 acres) which epitomizes the battleground at the urban–agriculture edge The land, once designated as ALR land and now in private ownership and out of the ALR, is the site of a proposed community with human-scale agri-food production as a central design element that will also be a significant economic driver in the proposed community (Southlands in Transition, 2009). In this project, theory and praxis literally meet 'on the edge'.

Conclusions

It seems inevitable that the concept of sustainability, in all of its dimensions, will come to define and focus human enterprise in the 21st century. The ideas offered in this paper are intended to stimulate creative thinking toward reconciling growth management, food security and the enhancement of agriculture. As the case study from Metro Vancouver illustrates, these are not three separate problems. Rather, they are multiple facets of the same problem. While we recognize that acceptance of our preferred vision may require a substantial paradigm shift, we are reminded of Albert Einstein's sage advice: 'The significant problems we face cannot be solved at the same level of thinking we were at when we created them'.

MEA represents a structured approach that can respond substantively to the economic challenges that will increasingly beset BC (food security defined in terms of supply, and food sovereignty defined in terms of control). We believe that it is important and timely to challenge the prevailing mindset that sees increased consumption as the measure of success, regardless of the implications on the resources depleted, ecological carnage created or the planning nightmares that are besetting our towns and cities. We depart from convention in suggesting that human-scale, agri-food production based on bio-regional rather than geopolitical boundaries represents an undervalued economic force that can transform how we design, plan and support our local communities. And we are working closely with a number of progressive municipalities throughout BC to explore how MEA can mitigate against the worst impacts of environmental and economic challenges, while at the same time showcasing practical ways through which to build the workforce (the next generation of urban farmers), the work (food security and agri-food production) and the productivity (urban agriculture as a significant municipal economic engine) that is essential for us to flourish.

References

Agriculture and Agrifood Canada, 2002, *Characteristics of Canada's Diverse Farm Sector*, Publication # 2109/B. Agriculture and Agrifood Canada, Ottawa, ON.

American Planning Association, 2007, *Policy Guide on Community and Regional Food Planning*, American Planning Association, Chicago, IL.

Archibald, H., 1999, 'Organic farming: the trend is growing!' *Canadian Agriculture at a Glance* 96-325-XPB, Statistics Canada, Ottawa, ON.

Artemis Agri-Strategy Group, 2002, *Economic Strategy for Agriculture in the Lower Mainland*, Metro Vancouver, Burnaby, BC.

Barlow, M., 2007, *Blue Covenant: The Global Water Crisis and the Coming Battle for the Right to Water*, McClelland and Stewart Ltd, Toronto, ON.

Baxter, D., 1998, *Demographic Trends and the Future of the Agricultural Land Reserve in British Columbia*, Provincial Agricultural Land Commission, Burnaby, BC.

Benbrook, C., Greene, A., 2008, 'The link between organic and health: new research makes the case for organic even stronger', *Organic Processing Magazine* March–April [available at www.organicprocessing.com/opmarapr08/opam08CoverStory.htm].

Benbrook, C., Zhao, X., Yanez, J., Davies, N., Andrews, P., 2008, *New Evidence Confirms the Nutritional Superiority of Plant-Based Organic Foods*, The Organic Center, Foster, RI.

British Columbia Ministry of Agriculture and Lands, 2006, *B.C.'s Food Self Reliance: Can B.C.'s Farmers Feed our Growing Population?*, Victoria, BC [available at www.agf.gov.ca/resmgmt/Food_Self_Reliance/BCFoodSelfReliance_Report.pdf].

British Columbia Ministry of Agriculture and Lands, 2008a, *British Columbia Agriculture Plan: Growing a Healthy Future for B.C. Families*, Victoria, BC [available at www.al.gov.bc.ca].

British Columbia Ministry of Agriculture and Lands, 2008b, *Planning for Profit: Five Acre Mixed Vegetable Operation – Full Production, Victoria, BC* [available at www.agf.gov.bc.ca/busmgmt/budgets/budget_pdf/small_scale/2008mixed%20veg.pdf].

British Columbia Ministry of Agriculture and Lands, 2008c, *Metro Vancouver Agricultural Overview*, Ministry of Agriculture and Lands, Victoria, BC.

Campbell, C., 2006, *Forever Farmland: Reshaping the Agricultural Land Reserve for the 21st Century*, David Suzuki Foundation, Vancouver, BC.

Carlson, A., 2008, 'Agrariansim reborn: the curious return of the small family farm', *Intercollegiate Review* 43 (1), 13–23.

Cavendish-Palmer, H., 2008, 'Planting strong boundaries: urban growth, farmland preservation, and British Columbia's Agricultural Land Reserve', Dissertation for Master of Public Policy. Faculty of Arts and Science, Simon Fraser University.

CBC News, 2008, *Canada's Inflation Rate Eases to 1.2% in December*, CBC News, Vancouver, BC [available at www.cbc.ca/canada/toronto/story/2009/01/23/inflationdecember.html].

City of Richmond, 2003, *Agricultural Viability Strategy*, City of Richmond, Richmond, BC [available at www.richmond.ca/services/planning/agriculture/viability.htm].

Condon, P., Mullinix, K., 2009, *Agriculture on the Edge: The Urgent Need to Abate Urban Encroachment on Agricultural Lands by Promoting Viable Agriculture as an Integral Element of Urbanism*, Institute for Sustainable Horticulture, Surrey, BC [available at www.kwantlen.ca/ish/urban.html].

Davis, D. R., Epps, M. D., Riordan, H. D., 2004, 'Changes in USDA food composition for 43 garden crops, 1950 to 1999', *Journal of the American College of Nutrition* 23 (4), 669–682.

District of Maple Ridge, 2009, *Maple Ridge Agricultural Plan Phase 3*, District of Maple Ridge, Maple Ridge, BC.

Ehrenfeld, J., 2009, *Sustainability by Design: A Subversive Strategy for Transforming our Consumer Culture*, Yale University Press, New Haven, CT.

Esseks, D., Oberholtzer, L., Clancy, K., Lapping, M., Zurbrugg, A., 2008, *Sustaining Agriculture in Urbanizing Counties: Insights from 15 Coordinated Case Studies*, University of Nebraska, Lincoln, NE.

Farmers Markets Canada, 2009, *National Farmers Market Impact Study*, Farmers Markets Canada, Brighton, ON.

Freyfogle, E., (ed.), 2001, *The New Agrarianism: Land, Culture, and the Community of Life*, Island Press, Washington, DC.

Garnett, T., 2008, *Cooking up a Storm: Food, Greenhouse Gases and our Changing Climate*, Centre for Environmental Strategy, University of Surrey, Surrey, UK.

Gillon, S., Minkoff, L., Thistlethwaite, R., 2006, *Grounding Ourselves: Innovative Land Tenure Models in California and Beyond*, California Food and Justice Coalition, Berkley, CA.

Goldberg, M., 2006, *Building Blocks for Strong Communities*. Research report F/58, Canadian Policy Research Networks, Ottawa, ON.

Goldschmidt, W., 1978, *As You Sow: Three Studies in the Social Consequences of Agribusiness*, Allanheld, Osmun & Co, Montclair, NJ.

Heffernan, W., 2005, 'Understanding what is happening in our food and farming system', in K. Mullinix, (ed.), *The Next Agricultural Revolution: Revitalizing Family Based Agriculture and Rural Communities, Proceedings of the Washington State Family Farm Summit*. Good Fruit Grower, Yakima, WA.

Heinberg, R., 2006, *Fifty Million Farmers*, E.F. Schumacher Society, Great Barrington, MA.

Hove, H., 2004, 'Critiquing sustainable development: a meaningful way of mediating the development impasse?', *Undercurrent* 1 (1), 48–54.

Illinois Local and Organic Task Force, 2009, *Local Food, Farms and Jobs: Growing the Illinois Economy*, State of Illinois General Assembly, Springfield, IL.

Ipsos Reid Public Affairs, 2008, *Poll of Public Opinion Toward Agriculture, Food and Agri-food Production in BC*, Ipsos Reid Public Affairs, Vancouver, BC.

Kaufman, J., Bailkey, M., 2000, 'Farming inside cities: entrepreneurial urban agriculture in the United Sates', Working paper WP00JK1, Lincoln Institute of Land Policy, Cambridge, MA.

Kent Agriculture Advisory Committee, 2004, *Small Lot Agriculture in the District of Kent*, Municipality of Kent, Kent, BC [available at www.fraserbasin.bc.ca/publications/documents/SLAinKentRptDec04.pdf].

Kimbrell, A., 2002, *The Fatal Harvest Reader: The Tragedy of Industrial Agriculture,* Island Press, Washington, DC.

Kirshenmann, F., 2004, *Are We About to Lose the Agriculture of the Middle?,* Leopold Center for Sustainable Agriculture, Iowa State University, Ames, IA.

Kirshenmann, F., Stevenson, S., Buttel, F., Lyson, T., Duffy, M., 2004, *Why Worry About the Agriculture of the Middle?,* Leopold Center for Sustainable Agriculture, Iowa State University, Ames, IA.

Korton, D., 2009, *Agenda for a New Economy: From Phantom Wealth to Real Wealth,* Berrett- Koelhler Publishers, San Francisco, CA.

McKibben, B., 2007, *Deep Economy: The Wealth of Communities and the Durable Future,* Holt, New York, NY.

Mendes, W., 2006, 'Creating and implementing a food policy in Vancouver, Canada', *Urban Agriculture Magazine* 16, 51–53.

Meter, K., Rosales, J., 2001, *Finding Food in Farm Country: The Economics of Farming in Southeast Minnesota,* Community Design Center, St. Paul, MN.

Metro Vancouver, 2007, *Census of Agriculture Bulletin #2,* Metro Vancouver, Burnaby, BC.

Metro Vancouver, 2009, *Metro Vancouver 2040: Shaping Our Future,* Metro Vancouver, Burnaby, BC.

Morton, B., 2008, 'BC farmers earning less: profits plummet as fuel and feed prices soar', *Vancouver Sun,* 26 May.

Mullinix, K., 2003, *The Next Agricultural Revolution: Revitalizing Family-based Agriculture and Rural Communities,* Good Fruit Grower, Yakima, WA.

Mullinix, K., Henderson, D., Holland, M., de la Salle, J., Porter, E., Fleming, P., 2008, *Agricultural Urbanism and Municipal Supported Agriculture: A New Food System Path for Sustainable Cities,* Surrey Economic Summit, Surrey, BC [available at www.kwantlen.ca/ish/urban.html].

Mullinix, K., Fallick, A., Henderson, D., 2009, 'Beyond food security: urban agriculture as a form of resilience in Vancouver, Canada', *Urban Agriculture Magazine* 22, 41–42.

National Farmers Union, 2005a, *The Farm Crisis: Its Causes and Solutions,* National Farmers Union, Saskatoon, SK [available at www.nfu.ca/briefs.html].

National Farmers Union, 2005b, *Solving the Farm Crisis: A Sixteen Point Plan for Canadian Farm and Food Security,* National Farmers Union, Saskatoon, SK [available at www.nfu.ca/briefs.html].

Patal, R., 2007, *Stuffed and Starved: The Hidden Battle for the World's Food System,* Harper Collins, Toronto, ON.

Penner, D., 2008, 'Land prices outstrip economics of farming', *Vancouver Sun* 24 May.

Pringle, T., 1994, 'Trust for BC Lands and Trust for BC Lands Foundation. Concept paper draft 2', Ministry of Environment, Lands and Parks, Victoria, BC.

Provincial Agriculture Land Commission, 2002, *Agricultural Land Commission Act* [S.B.C. 2002], Chap. 26, Provincial Agricultural Land Commission,Burnaby, BC [available at www.alc.gov.bc.ca/Legislation/Act/alca.htm].

Pynn, L., 2008, 'Farmland is like an endangered species: for farming to continue, BC must find ways for young farmers to make growing food products a viable career', *Vancouver Sun,* 24 May.

Quayle, M., 1998, *Stakes in the Ground: Provincial Interest in the Agricultural Land Commission Act. Part 4: Going Forward,* Ministry of Agriculture and Food, Victoria, BC.

Roberts, P., 2008, *The End of Food,* Houghton Mifflin, Boston, MA.

Rosenweig, C., Iglesias, A., Yang, X. B., Epstein, P., Chiuan, E., 2000, *Climate Change and U.S. Agriculture: The Impacts of Warming and Extreme Weather Events on Productivity, Plant Diseases and Pests,* Center for Health and the Global Environment, Harvard Medical School, Boston, MA.

Rowen, A., Duynstee, T., 2009, 'An analysis of the Agriculture on the Edge discussion paper, 2 April Agenda', Metro Vancouver Agriculture Committee, Burnaby, BC.

SmartGrowth BC, 2004, *State of the Agricultural Land Reserve in BC,* SmartGrowth BC, Vancouver, BC.

Smith, B., 2007, *A Work in Progress . . . the BC Farmland Preservation Program,* SmartGrowth BC, Vancouver, BC.

Southlands in Transition, 2009, 'Southlands: an open space neighbourhood', Post charrett paper [available at www.southlandsintransition.ca/].

Statistics Canada, 2008, *Study: Organic from Niche to Mainstream,* Statistics Canada, Ottawa, ON [available at www.statcan.gc.ca/daily-quotidien/080328/dq080328a-eng.htm].

Sustainable Development Commission, 2009, *Food Security and Sustainability: The Perfect Fit,* Sustainable Development Commission, London, UK.

Thompson, G., 2000, 'International demand for organic foods', *HorTechnology* 10 (4), 663–674.

Tunnicliffe, R., 2007, 'Saanich Organics: A model for sustainable agriculture through co-operation', *Columbia Institute for Cooperative Studies Occasional Paper Series* 2 (1).

United Nations, 2007, *State of the World Population 2007: Unleashing the Potential of Urban Growth,* United Nations Population Fund [available at www.unfpa.org/swp/2007/english/introduction.html].

Nourishing urbanism: a case for a new urban paradigm

Lewis Knight[1]* and William Riggs[2]

[1] Lewis Knight, Gensler, 2 Harrison Street Suite 400, San Francisco, CA 94105
[2] William Riggs, City & Regional Planning, UC Berkeley, 228 Wurster Hall #1850, Berkeley, CA 94720

True sustainability demands that we seek to more than 'prop up' traditional approaches to our environment; rather, it requires that we redress current shortcomings in the planning and design of our urban environment at both bio-regional and local scales. Nourishing Urbanism proposes a shift in the urban and non-urban paradigm relating to energy, water and food; all face significant climate-related challenges – and are united by land-use policy, planning and design. We need a renewed planning and design framework for cities and regions that allows the retrofitting of today's urbanity, and prepares our cities for a new tomorrow. Nourishing Urbanism seeks to provide a malleable planning and design framework that embraces the symbiosis between urban and non-urban, and provides for the well-being of the human condition through recommending policies and technical solutions that readdress land use, ultimately impacting the security of our energy, water and soil resources, as well as infrastructure, food supply, health and design.

Keywords: architecture, climate change adaptation, land planning, planning theory, urban agriculture, urban and regional design

Introduction

It was with a sad realization that an 8-year-old girl returned from school confused. She was one of a few children in her class who believed an apple was picked from a tree. Many in her class of budding environmentalists and professionals appeared unaware of the idea of rural and urban interreliance. They believed an apple came from a box at the local supermarket! Luckily, trips to visit rural family had 'fed' her the knowledge that her urban life depended on its relationship with the rural.

Increasingly, the urban (city condition) is isolated from the non-urban (rural, and other), yet ongoing research and documents such as the Millennium Ecosystem Assessment prove their symbiosis is vital to the success of both. The goal of this article is to review the apparent isolation between the urban and non-urban, discuss the trend toward a greater urbanization and its implications for both urban and non-urban constructs, and propose land-use planning and design change.

This article will review urban change and the increasing dominance of the urban condition and, as it becomes more prevalent, how the urban tends to cannibalize the non-urban, resulting in potential failure of both. Utilizing the notion of a basic (biophilic) human need to connect to the natural environment, the article will provide a structural and theoretical basis on which to address issues including energy, water, soil and urban agriculture in a 'nourishing' fashion that reinvents the traditionally separated urban–rural land-use paradigms to promote a more transparent, enriching and ultimately resilient relationship of commonality between the urban and non-urban. The objective is to leverage the dominance of the urban to positively influence its own strength and resilience, and use this dominant role as an agent to reinforce and strengthen the non-urban environment.

The current paradigm

For many, the paradigm of incompatible urban and non-urban environments is an inescapable and divisive reality. It is reinforced by economies, governments, cultures and history. From the earliest collection of families

*Corresponding author. Email: lewis_knight@gensler.com

INTERNATIONAL JOURNAL OF AGRICULTURAL SUSTAINABILITY 8 (1&2) 2010

PAGES 116–126, doi:10.3763/ijas.2009.0478 © 2010 Earthscan. ISSN: 1473-5903 (print), 1747-762X (online). www.earthscan.co.uk/journals/ijas

forming small encampments and villages, to our mega-cities of today, the relationship between urban and non-urban has frequently been cast as adversarial. This has not always been the case. An exception to this may be found in the city-states of the Italian Middle Ages. These communities are an example of a more balanced and synergistic relationship between the urban and the non-urban. The non-urban was able to substantially feed the city at its heart. The city provided the opportunity for government, religion, economies and cultures to flourish through interactions, not only inside the city and with other cities, but with the non-urban. However, while this symbiotic relationship between urban and non-urban defined cities in our past, it does not frame our present global urban convention of consumption of non-urban land by the urban.

Contemporary development patterns

During 2008, the world became largely urban; more than 50 per cent of our six billion population now live in cities. While this is familiar for some, it is a new phenomenon for most (Burdett and Sudjic, 2008). We are part of a great urbanization that shows few signs of slowing. Most global population projections agree that at the end of the 21st century there will be between eight and 15 billion people on earth, after which population growth may plateau. Of this anticipated growth, the majority will be likely to be absorbed in cities, both existing and new.

This urbanization is a good thing; it promotes interactivity, education, social advances and global human awareness and equity. However, it presents serious environmental challenges that require innovation. As with the demands for urban space, pressures on the environment are also creating a unique paradigm that urban and non-urban are divided despite being coexisting ecosystems. This division between urban and non-urban is of particular concern with regard to overarching environmental issues including climate change, water rights and energy policy.

Development patterns and growth around cities follow a familiar trajectory. In 19th-century London, the planned 'garden cities' became peripheral refuges for the upper and upper-middle classes who commuted to London for work. 'Metroland' was successful as many fled the moral and physical diseases (miasmas) of the inner city for the beauty and purity of the country. Yet the success of these suburbs pressured the greenbelts that were part of their original concept (or design intent) and were intended to connect town with country; and provide all the benefits of country.

After World War II, community design and expansion continued, most frequently outside existing city boundaries (Hall, 1996). Housing construction was encouraged to boost the economy and the transition from military production. These post-war suburbs were populated homogeneously with the middle classes. In the United States, places like Levittown, ex-urban suburbs dependent on private automobiles, became the new urban model (Jackson, 1985). Transportation and commuting patterns reduced the functions of existing and new towns; many became bedroom communities, creating little connection between either urban or rural.

This 'sub'-urban framework produced developments with large-lot homes and dispersed neighbours. While garden suburbs were found to have greater neighbourhood involvement than their urban counterparts, they were largely filled with homogeneous populations which, over generations, created difficulty in social connectedness (Putnam, 2000). Occupants seeking privacy and safety found isolation and social disconnection. Evidence suggests this leads to unhappiness because of the loss of the social capital found in traditional communities.

These 'sprawling' developments have little connectivity on foot. The ease of non-motorized travel to schools, stores and workplaces is limited (Sallis et al., 2004). Further, these activities are often unsafe in communities that were planned for vehicular dominance and built for inactivity. Statistically, suburban residents are more likely to have chronic mental or physical health conditions (Sturm and Cohen, 2004). These development patterns have also contributed to poverty and blight, exemplified in the United States by minorities who are unable to find adequate or healthy housing in inner cities, yet who are unable to afford suburban homes.

This trend away from a more integrated urban model – with interdependence between urban and non-urban – has been recognized as a global phenomenon. It is not one limited to the developed world, and has led to new phrases such as 'edgeless cities' (Lang, 2003). Edgeless cities lack diversity in land use, have inadequate or inaccessible open space, provide limited opportunity for local food production and frequently isolate housing from jobs by not providing transit.

Whether or not sprawl is low density or uncoordinated high density (such as East Asia), the underlying assumption is that, as a phenomenon, it embodies inefficient use of land as a resource. The result continues to be communities that suffer from 'spatial mismatch' and are isolated from the services that support them, placing an unfair burden in allocating resources (environmental and other) on a regional level.

New Urbanism and Smart Growth (Orinco Station in Portland, Oregon; Poundbury in Dorchester, England; Breakfast Point in New South Wales, Australia) are examples of trends seeking to address issues of ex-urban growth and dispersed urbanism. Both borrow lessons from historic successes and employ more traditional neighbourhood patterns, along with parks and boulevards reminiscent of 19th-century design. However, these 'movements' have largely maintained the pattern of market-based planning and urbanism and, despite claiming to do so, have neither significantly enhanced urban social or environmental equality at a regional scale nor truly provided a 'nourished' urbanism that focuses on the preservation and integration of the non-urban framework.

Whether in the Americas, Europe, Asia or South America, urban success typically leads to horizontal expansion of the city, most often consuming valuable agricultural land at a city's margins to do so. Our challenge is to define an urban paradigm that is healthier, protects our land-based assets more strongly, and supports the ongoing growth and health of our cities.

Global warming and environmental stewardship

Global warming is one of the key issues that must drive the need to reframe the urban–non-urban paradigm. *An Inconvenient Truth* raised awareness of global warming at a time when the Kyoto Protocol was not recognized by the incumbent US administration. Yet Al Gore's message requires amplification. MIT researchers estimate that median global temperatures will rise 5.2°C by 2100, double earlier estimates (Sokolov *et al.*, 2009). Three primary issues challenge the resilience of our cities and require assertive environmental stewardship that nourishes urbanism: climatic volatility; sea-level rise and flooding; and change in production systems and methods. During the 2009 bushfires in Victoria, Australia, the ABC reported significant livestock losses. Simultaneously in Queensland, flooding led to herd losses of as much as 50 per cent. While these events are not solely attributable to global warming, they provide a clear indication of potential stock, field and food production losses due to a more volatile climate. This climatic volatility has the potential to produce a more volatile food supply network, with the consequence of greater social and economic instability.

According to the Pew Center on Global Climate Change, flooding at catastrophic levels and sea-level rise has the potential to cause significant inundation of our cities, coastal wetlands and low-lying agricultural lands. The threat of sea-level rise, already being experienced on islands in the Pacific and producing 'environmental refugees', is a significant threat to sustainable agriculture, as many deltas are home to major agricultural production systems. Global warming has the potential to drive significant migration of agricultural production in non-urban areas to portions of the globe once considered unfeasible for crops, thus reinforcing the idea that the urban must carry at least some of this burden. According to a 2007 study, warming temperatures may have reduced the combined production of wheat, corn and barley by 40 million metric tons per year between 1981 and 2002 (Lobell and Field, 2007). The study estimates the annual losses at $5 billion. Apart from making a case for the long-term impact of warming on traditional crops, this provides an insight into the potential need to modify crop selection as well as migrate production areas to climatic zones that will, in the future, provide better growing conditions.

An issues-based framework

While global warming presents an agenda for many, little climate change research investigates the ramifications for land-use planning and design. The interdisciplinary nature of planning and design can, however, become a bridge to environmental activism for both the urban and non-urban.

Energy

Of the major energy users in land-use terms, transportation and buildings account for the majority of consumption. Adapting to carbon-free energy is critical to defining a resilient future for our cities, both in terms of consumption reduction and migration to true renewability of energy sources. An understanding is evolving of the relationship between transportation infrastructure and land use, and the impact on energy consumption and efficiency, with vehicle miles travelled (VMTs) becoming an index for both.

There has been major change in our product supply network since WWII. Stemming from technological advances and the adoption of the road-rail-ship container, our food industry has become more global, more centralized and more concentrated (Oatkiss, 2005). In the UK, food transport accounted for almost 30 billion vehicle km in 2002, producing 8.7 per cent of vehicular CO_2 emissions. In the United States, it is estimated that the average grocery store item travels almost 1500 miles (2400 km) between farm and refrigerator (Thomas and Drukker, 2009).

Oil, may, in the short term, be replaced by other carbon-based fuel sources, such as coal, natural gas or shale oil that require extraction, or biofuels that require a new balance of land use in non-urban environments. In the longer term, alternative technologies that address climate issues must be allowed to develop.

Irrespective of the replacement of oil, for our urban future to be resilient, we must design for a more frugal short term that protects our environmental assets, while maintaining adaptability to future technologies. The burden of the planning professions is to reduce energy consumption through more compact urban form, while also promoting energy innovations that may require new planning methods.

Water

Water is also a carrier of disease and pollution, and a resource that is at risk on a global scale. More than 40 years ago, Rachel Carson warned in *Silent Spring* of the cumulative impact of toxins in our environment; in many cases these were waterborne (Carson, 1962). Recent studies have linked build-ups of various human-made contaminants (largely pharmaceuticals) with deformed sex organs in Florida alligators, and with polar bears in the European Arctic becoming more hermaphroditic. Equally crucial, as it affects our food systems from fish to larger mammals, is the ability to sufficiently clean pollutants so they do not contaminate food sources (Clover, 2006).

The World Health Organization estimates that almost 884 million people lack access to safe drinking water (United Nations, 2009) while 2.5 billion have no sanitation, leading to increased incidence of disease. In the developed world, declining infrastructure expenditures and facility failure place strain on existing water supplies and sanitation.

Nevertheless, water infrastructure is a significant contributor to the built form of cities. Many of the great cities of the world are located on a waterway, be it an ocean, sea, lake or river. Where water is scarce, as in Bangalore (India) and Las Vegas (USA), growth of the contemporary city is paralleled by construction to ensure water supply. In both cases, the underground aquifer is challenged to meet demand and, in recharge, clean water is often replaced by a more polluted equivalent.

Water infrastructures have formed the basis for major urban renovations within the city. The canals of Amsterdam and St Petersburg were developed to enable city construction. Both the 'modernization' of Paris under Baron Haussman in the mid-1850s, and the construction of London's Embankment during the 1860s,

exemplify the pursuit of clean water and sewage leading to great city-making ventures.

Ironically, while our cities are sources of significant point-source pollution, they also have sufficient population and centralization to make them the ideal treatment centres for our waters, both fresh and saline. In essence, our cities will need to adapt to a more systemic relationship with water, becoming the filters of water in the environment, rather than merely consumers then dischargers.

Soil loss and nutrient depletion

Loss of arable land to production is a significant risk to agriculture systems. In China, 10 per cent of arable land available in 1979 has been lost to rapid urbanization (McKinsey Global Institute, 2009). Should this trend continue, by 2029 a further 15 per cent will be lost to urbanization, or almost 20 million acres. Much of China's urban expansion has occurred in the relatively rich grain bowl in the southeast of the country, potentially exacerbating the loss of high-yield arable land. The Chinese government has identified the loss of productive land as a significant issue, and since 1986 has trialled several mechanisms to maintain minimum arable land values in the nation. The success of these moves will be vital to the country's future as it continues to become more urban.

Loss of soil fertility presents a similar issue. In large part, this is occurring across the globe where new populations either mismanage or deliberately over-produce. For the first time in its history, Egypt has become a net importer of food; it is no longer able to feed itself, as the Nile Valley has become less and less fertile (Montgomery, 2007).

Mismanagement of our productive landscapes, both urban and non-urban, is also a significant issue. Declining agricultural yields, competition for crop space, and continued release of carbon dioxide into our environment due to deforestation all illustrate the misuse of land as a resource.

While other reasons exist for the loss of arable land, historically, the more successful the city, the more voracious its appetite for land. Most typically, the most successful cities are located nearest to their food source and its most fertile soil assets. At issue is a truly sustainable future for ourselves and our planet, one that is not reduced to an equation, and one that recognizes the positive power of the human condition.

Food

Urban food production is a feature of many cities. In Havana almost 90 per cent of food consumed is produced within the boundaries of the city (Viljoen, 2005). The same occurs in Shanghai where the informal sector occupies vacant land to provide a significant portion of the city's daily greens.

Urban food production is not solely a developing world occurrence. Allotments in London arose under the auspices of the House of Lords and were seen as a way (in perpetuity) of attempting to supply cheaper foods for the working class. In Boston, as in London, urban farming developed on a more formal basis when parks were converted to allotments between the World Wars. Urban allotments and gardens have provided a stable yet flexible supply framework, but are now under ever greater pressure of displacement by development, and require greater protection (Viljoen, 2005).

In the United States, local food, slow food and community-supported agriculture are gaining social momentum, and offer opportunities to reinvent a combined urban and non-urban paradigm. However, most focus on 'in-season' food produced on smaller scales without the aid of pesticides or chemicals. The Greenbelt Alliance in San Francisco advocates expanding local food production to the regional park system. These solutions, with others, are being tested by organizations such as the UC Davis Institute of Sustainable Agriculture. There is, however, growing consensus that practices that reduce food miles contain significant environmental benefit. They also support biophilic concepts and can help beautify, enrich and educate urban lives.

Reinstating the criticality of urban agricultural space to feeding and nurturing the public is a key to confronting the growing distance food travels. Alongside a re-education of diets and palates, the environmental and health benefits of eating vegetables and low-food-chain, limited off-gassing meats such as sustainable fish, and the environmental costs of non-urban commodities (fuels, clothing, and building materials) are an integral part of a new urban paradigm that requires responsible land-use policy and planning.

Biophilia (and the human condition)

We are now faced with a reality in which we are more disconnected and distant from the natural environment within the urban framework, yet we are finding more need for humans to have a connection to the natural world. Roger Ulrich is convincing in his studies of hospitalized patients (Ulrich, 1984). Using the control of a hospital environment, he studied patient recovery rates for those with and without contact to the natural environment. Patients with rooms facing a park had a 10 per cent faster recovery and needed 50 per cent less pain-relieving medication when compared to patients in rooms facing a building wall.

Subsequent research involving those exposed to stress-inducing environments – including hospitals, prisons, offices, military camps and horror films – has indicated in many cases that those with access to greenspace and views of nature have reduced stress reactions (Ulrich et al., 1991). When subjects of such experiments were exposed to natural environments their levels of stress decreased rapidly, whereas during exposure to the urban environment their stress levels remained high or subsequently increased.

While some researchers reference this synergy between the built environment and health as a 'biophilia' hypothesis – the belief that there is an inherent need for connection with nature in every human being (Frumpkin, 2003) – others point to scientific research that concludes that when individuals think about the natural environment the brain is relieved of 'excess' circulation (or activity) and nervous system activity is reduced; thus, stress is relieved (Maller, 2005).

Some contest human reliance on the natural world based on factors such as race and socio-economic status; however, many professions recognize the importance of 'genus locii', 'sense of place' and 'physical beauty'. These require aesthetic or architectural frameworks (including impacts of spatial orientation, memory, passion, and sacred or social constructs) to connect humans with the natural environment. Some studies in the field of psychology suggest that we need to interact with other organisms in addition to the natural environment to reach maturity; and that lack of interaction with nature (living in a 'denatured environment') leads 'to a society of childish adults' (Dekay and O'Brien, 2001).

Land use and buildings as the keystone

In concurring with 'biophilia' there is a need to emphasize the connection between the built environment and the natural world. However, much urban space continues to be designed and planned, particularly in the developing world, with a consumptive attitude to the non-urban landscape. The idea of integrated natural beauty for agricultural utility and for biophilic benefits is muddied by scientific and sustainability metrics – the engineering-based concept that we should be able to quantifiably measure the benefit of environmental impacts. Correspondingly, the notion of natural beauty for beauty's sake is largely dismissed. Better building science that integrates urban agriculture, sustainable transportation and active living environments can help, but the pursuit of natural beauty within the urban framework is equally important, and should become part of the overarching framework by which we measure the success of our cities.

Many studies suggest that urban access to food systems and the natural environment has been compromised. In Los Angeles during 2006 the Rand Corporation found that the majority of the city population was underserved by the park system, and did not have sufficient access to parks and open space to maintain physical activity (Cohen *et al.*, 2006). The study showed that most parks were more than three miles from the residents they served, becoming essentially unusable. When recreational assets at local churches or school yards were close, they were frequently locked and inaccessible during the evenings and weekends when usage would be higher (Scott *et al.*, 2007).

We now understand that connections between land use, transportation and the environment play a large role in food access and a healthy urban experience (Frank, 2000). The act of driving, largely contributed to by urban development patterns, causes stress and mental fatigue (Frumpkin, 2002), and increased driving to buy or transport food and to get to work or play has been correlated to increased in body mass index (BMI), equating to a six per cent increase in the likelihood of obesity (Frank *et al.*, 2004). Further, there are generations of US citizens who face location-based discrimination in their ability to access healthy food (Morland, 2002). Clearly, it makes sense to rethink land use and the urban framework to reduce driving trips and increase access to the natural environment.

If simply 'greening' environments can be both healthy and have climate change impacts, a new emphasis on urbanism that integrates high-performance, high-quality green space within the built environment must provide a mechanism for improving our cities. This forms the basis of suggested new directions and a theoretical framework that fuses the urban and non-urban. By nourishing urbanism with the pragmatism and beauty embedded in the natural world we can begin to reinvent and reshape the urban and non-urban paradigm as it currently stands. It calls for practitioners to embrace the non-urban environment as a part of the urban when thinking about the future.

What is the recommendation for our urban context?

While issues of transportation and the market economy play roles in shaping urban agriculture and open space,

there are critical land-use planning, urban design and architecture opportunities that can provide the backbone for nourishing urbanism and address many of the previously identified issues for environmental activism. These opportunities form a baseline for reinventing the urban and non-urban paradigm – a reinvention that starts with some of the basic building blocks of our society, or how we build and develop our land.

Energy

- *Reduce food miles.* Efforts should be made to reduce both miles travelled and energy consumed in feeding ourselves. We must be careful that the availability of diverse diets does not continue advancing at the cost of the environment. While biofuelling and alternative vehicles can provide some gains in this arena, a policy which requires a labelling scheme identifying the distance food has travelled and its carbon footprint could reinforce the value (both environmental and market) of local food. This, combined with more efficient transit and a revisitation of how urban land is used for agriculture, could create dramatic reductions in food miles.

- *Develop alternate (synthetic) materials suitable for fuel, housing and clothing.* Material science will become more essential as we continue to degrade our soils – irrespective of the ability of production land to keep pace with the demand for these materials. Additionally, there is a new generation of synthetic fabrics that are intended to be biodegradable. Many, while artificial, are made from agricultural by-products.

- *Incentivize land-use policy that increases access to public transport.* Increasing land-use mix to encourage 'active transportation' for trips to local food and services that increase physical activity (Sallis *et al.*, 2006), and encouraging shorter blocks, smaller streets and higher density (Frank, 2005). Density alone is proven to be one of the strongest indicators of walking.

Orinco Station in Portland, OR and Atlantic Station in Atlanta, GA integrate features associated with healthier residents at work and play to encourage walking. These include: increased mix of accessible uses (residential and at least one other use); safety elements such as wider or more sidewalks, pedestrian refuges and traffic-calming measures; providing lanes and parking for bikes; 'more windows facing the street and more street lighting, and fewer abandoned buildings, graffiti, rundown buildings, vacant lots, and undesirable land uses' (Alfonzo *et al.*, 2008, p. 44); reduced incentives for driving such as free parking and using the private market to set a value on common resources such as air quality.

Many communities are now using location-efficient loans to encourage greater levels of density in priority development locations, incentivized through reduced interest rates to consumers. In places like the San Francisco Bay Area these loans are being discussed under the regional 'Transportation for Liveable Communities' scheme. Combined with local funding for street-level pedestrian safety improvements, these programmes should increase transit access, not only for transportation purposes but also for the distribution of agricultural and commercial products.

Water

- *Plan for shortage (and tell the story)*. Greater emphasis should be placed on water as a limited and/or polluted commodity. This must be evident in building codes and development, and must become a feature of the built environment showcasing it. Many architects and planners are showcasing seasonality in the urban landscape, rebuilding/recreating traditional marshes and wetlands, and building networks of 'waterscapes' that restore natural watersheds, passively improve water quality and educate the public about ecosystems. Requiring such features in building codes would create a greater awareness of 'living' water features that are tied to the land and seasonal changes in water supply.
- *Footprint all development*. Water footprinting should be embraced on a wider scale in development evaluation, and is likely to be used more frequently in California under a recent Assembly Bill. It is proving a reliable method for evaluating global water use. However, trade should be a significant consideration as many water-rich environments bear significant unaccounted burdens by exporting to more arid environments, paying the burden for increased consumption and pollution that occurs as the producing nation (Hoekstra and Chapagain, 2008).
- *Zero pollutants*. Like calls for zero-carbon or waste-free communities, zero waterborne pollutants must be a goal of all urban areas. Our cities have the resources to clean fully and must do so, to protect downstream populations and environments.
- *Mandate on-site water retention and preservation*. Options include urban marshes, aquifers and increased water storage and retention areas. In Australia, water-sensitive urban design and in the United States, low-impact development integrate stormwater management and treatment into the urban environment. In

the UK, *Climate Change Adaptation by Design: a Guide for Sustainable Communities* recommends planning mitigation for future changes in water, including regional interventions that include: water storage in existing aquifers; flood attenuation; increased tidal defences and sea walls; and run-off management. Much of this strategy is articulated in projects such as the *Adaptable Urban Drainage Project* (available at www.k4cc.org/bkcc/audacious).
- *Mandate 'clean' aquifer recharge policies*. Like many extraction industries, removal of water from aquifers needs to be balanced with clean water recharge. Declining aquifer levels and increased pollutants are causing significant issues globally.

Soil

- *Entitle soil nutrient recharge*. Divert waste, and establish nutrient replacement cycles from the urban to the non-urban through both policy and land-use planning. This will potentially be very valuable where urban agriculture is fully integrated into the urban land-use pattern, and is already being developed by the East Bay Municipal Utilities District in Oakland, California, where organic industrial and agricultural waste is being accepted to balance the waste stream, drive an energy plant, and provide clean compost for re-use on agricultural lands.
- *Establish valuation of lands that favour long-term productivity*. Establish price controls and preservation targets that encourage and support productive land preservation and incentivize life long soil replenishment and management. One way of doing this could be to develop new/more broadly defined economic indices that prioritize net future value and incorporate evidence-based analysis of healthy lifestyles. By incorporating these factors into the economic modelling of development, a more far-reaching understanding of development patterns should be incorporated into urban infill and new developments.

Food

- *Promote regionally based design*. Adopt planning and design that is critically regional. Unfortunately, not all environments are equal, and a regional perspective is essential to balance resources and provide for the common good. Local environments require the development of critical responses that are tailored to their situation, rather than the application of normative planning and design principles. Each city region should be planned as a whole, protecting food sheds, reducing vehicle miles and generating a balanced attitude to land use and equity.

Agriculture is typically practised on the outskirts of cities in planned greenbelts that surround cities but do not always best serve them. Sweden and Denmark have provided examples of radial areas of greenspace extending from a town centre, connected by rail, bike and pedestrian routes between Malmo and Copenhagen, but this greenspace has focused on space solely for recreation rather than for productive agricultural uses (Hall, 2009). Oft-cited Village Homes in Davis, CA integrate more agricultural space, yet leave out transit infrastructure in favour of the automobile.

A true complete integration of the urban and non-urban should be pursued as a new theoretical framework in the planning, architectural and engineering fields, recognizing a role for the rural greenbelt but integrating non-urban uses within the urban core. Some of these original concepts of 'radiating' greenspace throughout the city lie not in the theory of greenbelt preservationists like Ebenezer Howard and Frederick Olmsted but in modernists like Le Corbusier.

New 'ecotown' concepts planned as new sustainable communities in the UK claim to achieve this, but paradoxically are planned on 'greenfield' sites formerly used for agriculture and grazing. 'The Preserve' in Stockton, CA seeks to provide agricultural land throughout the 1800-acre development which balances water usage and seeks to restrict VMTs.

- *Encourage new forms of land preservation and conservation.* Through promoting innovative management and design of publicly held land and open space. Berkeley, CA is working with landscape designer Walter Hood to create green, open space corridors through its downtown, and New York City has converted the former High-Line Rail into an elevated urban park. These spaces provide integrated natural space and define a community identity which educates locals about the origins of their food, water, supplies, and their unique place in the natural ecosystem.

- *Embrace innovative approaches to increased density that support land preservation, conservation and management for agriculture and open space.* Through an integrated regional land-use approach aimed at providing a platform for regenerating and improving the urban and non-urban framework. Examples such as the concept plan for new development at the Napa Pipe project in Napa, CA (near protected Napa Valley wineries) illustrate how new development can knit the urban and non-urban. Homes maximize land-use efficiency to provide significant communal space in the form of shared green open spaces and the preservation of agricultural land; however, even these examples may not go far enough towards achieving true symbiosis between urban and non-urban.

- *Establish policy to produce and grow food locally on public land.* Many US cities are establishing mandatory urban agriculture programmes that take advantage of vacant lots, rooftops, medians and public open space for food production and education. San Francisco has sent model ordinances to city legislators suggesting that food be produced within the city's border and that city agencies source healthy food from within its foodshed. The city allows public food subsidies to be accepted at farmers' markets.

- *Establish minimum maintenance standards for undeveloped land.* Land banking and ownership should require a minimum standard of maintenance or production, so that absentee landlords manage land in an appropriate fashion for the broader ecosystem.

- *Consider whole-life management practices for public facilities such as schools.* Food and nutritional policies should encourage community partnerships, local and organic production, and the requirement of schoolyard gardens to produce organic produce. School programmes are already in place and have been used in Berkeley, San Diego and Philadelphia, positively impacting on educational environments and bringing healthy food options to local schools (Center for Health Improvement, 2006). Although more research is needed, there is anecdotal evidence that such programmes can increase student health and academic performance (Ozer, 2007).

- *Promote urban research that verifies innovative mixed-use such as 'vertical agriculture'.* Vertical agriculture is becoming more viable and cost-effective. It increases the possibility that buildings and agriculture can symbiotically coexist – and opens the possibility that the urban provides not only jobs and housing, but feeds, clothes and nurtures the health of residents.

During a recent exhibition at Ryerson University in Toronto, the prospects of urban agriculture were explored by other disciplines (Carrot City, 2009). Dutch architectural firm MVRDV provided a vertical Pig City, while others explored the possibility of vertical agriculture both as integrated in the built form of skyscrapers, and as a new typology of vertical greenhouse. While potentially more far-reaching in vision, there is significant evidence that these structures may prove a viable form of providing food to the urban some time in the future.

Address social equity connections to food and open space resources. Most literature recognizes access to healthy food as an index for decades of segregation and inequality (Williams and Collins, 2001). Socioeconomic status should not mean that one lives without a nutritionally balanced diet, has no access to public and private transportation, is exposed to higher crime, has less retail amenity, and lives with problems such as litter, noxious odours and discarded needles.

The Robert Wood Johnson Foundations' Active by Design programme provides an example that has supported noteworthy, community-based health programmes in cities such as Boulder, CO, Portland, OR, Cambridge, MA, Olympia, WA, and Lexington, KY that have worked to establish healthy habits in local residents (ALBD, 2006). These programmes recommend an ecological approach to chronic diseases such as obesity that focuses on multiple pathways to disease – addressing the built environment at the same time as behaviour. Programmes that embrace this balanced approach should be used to support more equitable access to resources.

Educate about healthy diet. Diet as a factor relates not only to urban agriculture but to beauty and enrichment for our largely sedentary urban lifestyles. Children in the United States spend over four hours per day in front of the television (Robinson, 1998) and 31 percent of the adult population is obese (American Obesity Association, 2008). Education on nutrition and healthy, low-impact eating should come at the earliest periods, creating sustainable behaviour that lasts a lifetime.

Policies that focus on food preferences have been ineffective in fighting addiction to diets heavy in artificial sugars, large amounts of high-fructose corn syrup (HFCS) and saturated fats. More aggressive policies should be pursued such as the taxation of sodas and junk food, and the clear labelling of sugar additives such as HFCS. Money from these taxes could then be used to subsidize schoolyard gardens, walking to school programmes such as the International Walk to School Day (www.walktoschool.org/) and to turn parking spots into green spaces, as happened in the recent PARKing Day (www.parkingday.org/).

Infrastructure

- *Intelligent and unique infrastructure.* Roads and urban infrastructure are not specifically part of this article. However, infrastructure, particularly below-ground utilities and roads, is one of the most resilient contributors to urban form. Historically, it has provided an equitable supply of commonly needed resources. In order to manifest Nourishing Urbanism, infrastructure will need to be modified and/or reconfigured to support actions described elsewhere.
- *Promote multiple-use/recycling/repurposing of structures.* Review building codes which separate uses, and require access and egress that is different for different uses. The elevator has become the freeway of the vertical city – isolating and antisocial. A review of all building codes is required to make construction both simpler and also more adaptable to alternate uses, and more able to accommodate future unforeseen uses such as vertical agriculture.

Design

- *Design is important.* Promote a culture of design excellence. Many cities are beginning to realize the economic and social benefits of a well-designed contemporary urban public realm. Frequently, when confronted with social change, communities opt for historicist design responses (new urbanism/neotraditionalism), yet contemporary urban society requires innovation in the urban environment.
- *Develop social indexes as measures.* Increasingly, planning will need to be more than performance-based; it will need to become evidence-based. Clear indicators against which long-term performance is measured will need to be determined beyond economic productivity of a city or region. Ironically, obesity levels are rising in the United States, China and India. Diet is often cited, yet there is also correlation between increased dispersion of urban services and health concerns. Increased community health, longer life spans, less expensive (and more preventative) medicine, and higher academic achievement will all be achieved, should planning embrace whole-life systems rather than static horizontal consumptive models. These indices must provide a more compelling measurement of the way we plan.
- *Research and design new building technologies and materials.* Locally source sustainable materials to aid in housing our growing population. These technologies should be offered within the framework of a holistic design that integrates the non-urban and urban ecosystems while being well designed. Many ratings frameworks (LEED, BREAM, Green-Globes, Build-It-Green) attempt to address sustainable materials but none achieve an 'ecosystems'

ethos that integrates urban and non-urban. Green building standards should enliven the built environment but not prohibit density and creative, healthy and beautiful urban spaces.

Using observational evidence from other projects, hospital environments are being designed to integrate natural views to improve patient, visitor and staff experiences and connect them to nature. This trend is finding traction in the planning and design community in London through the work on Space Syntax by Bill Hillier, and deserves greater exposure.

- *Provide increased access to parks and open space.* Access to usable and meaningful open space should be a priority, not only through exactions on developers, but must be incentivized. Despite the environment having health benefits and prolonging lifespan, many communities still have little green or natural space accessible to residents (Takano *et al.*, 2002). There are limited national or international benchmarking standards for local open space policy. In the past a rule of thumb has been 1 acre per 1000 persons; however, few if any cities achieve this.

- *Develop operations and management mechanisms that are flexible.* Accessible open space in the urban setting is essential for human life and should be addressed directly by international benchmarks and policy at the regional level. Allocations could be made for regional open space accessibility – allowing for all populations to have equal access to healthy lifestyle choices and active spaces. These targets could in turn be articulated to regional, state and national decision makers in the hope that the *entire* urban environment could be 'nourished' through regionalism and intergovernmental cooperation.

Conclusion

We need a new paradigm; a construct that reinvigorates the urban and the non-urban, which evaluates and reframes the professions responsible for planning and design of our environments. We must transform our traditional professional roles and responses. It is time for a new normative framework that turns environmental advocacy into activism. We must integrate the personal and the professional in order to nourish urbanism and hence nourish ourselves and our families. We must not only develop a new and more nuanced way of designing environments that reaches further than previous professions; we must conduct our professional practice, living, working, playing, eating, gardening and greening our immediate environments.

There is a need to reconnect our urban and non-urban environments through a reinvigorated urbanism that fully embraces the notion of healthy environments. Urban land should be fed and, in turn, should nourish the urban condition. It should not only feed us, but clothe us, house us and fuel our mobility with goods, services and agriculture. It should complement and provide beauty and meaning to our time spent walking to school, socializing at friends' houses, going to the library, the store or the park. It should not be sustained as our term 'sustainability' would imply; it should be 'enriched' so that the benefits of the natural world can be felt by all, regardless of age, race, class or creed.

In doing so, we can ensure that our children are not confused about where their food comes from. They will know that apples don't just come from boxes in stores; they are grown in schoolyards, backyards, streets, and unique non-urban, natural places that nourish the urban environment.

References

ALBD (Active Living by Design) Case Studies, 2006, *Robert Wood Johnson Foundation* [available at www. activelivingbydesign.org/index.php?id=342].

Alfonzo, M., Boarnet, M. G., Day, K., McMillan, T., Anderson, C. L., 2008, 'The relationship of neighborhood built environment features and adult parents walking', *Journal of Urban Design* 13 (1), 29–51.

American Obesity Association, 2008 [available at http://obesity1. tempdomainname.com/subs/fastfacts/obesity_US.shtml].

Burdett, R., Sudjic, D., 2008, *The Endless City*, Phaidon, London.

Carrot City, 2009, *Carrot City On-line Guide*, Ryerson University, Toronto, Canada [available at www.verticalfarm.com].

Carson, R., 1962, *Silent Spring*, Houghton Mifflin, Boston, MA.

Clover, C. Ch., 2006, *The End of the Line: How Fishing is changing the World and What We Eat*, University of California Press, Berkeley, CA.

Cohen, D., Sehgal, A., Williamson, S., Sturm, R., McKenzie, T. L., Lara, R., Lurie, N., 2006, *Park Use and Physical Activity in a Sample of Public Parks in the City of Los Angeles*, RAND Corporation, Santa Monica, CA [available at www.rand.org].

Center for Health Improvement, 2006 [available at www.cahpf. org/GoDocUserFiles/130.Case%20Studies%205.3.06.pdf].

Dekay, M., O'Brien, M., 2001, 'Gray city, green city', *Forum for Applied Research and Public Policy* 16 (2), 19–27.

Frank, L. D., 2000, 'Land use and transportation interactions: implications on public health and quality of life', *Journal of Planning Education and Research* 20, 6–22.

Frank, L. D., Andresen, M., Schmid, T., 2004, 'Obesity relationships with community design, physical activity, and time spent in cars', *American Journal of Preventive Medicine* 27 (2), 87–96.

Frank, L. D., Schmid, T., Sallis, J., Chapman, J., Saelens, B., 2005, 'Linking objectively measured physical activity with objectively measured urban form findings from SMARTRAQ', *American Journal of Preventive Medicine* 28 (2).

Frumpkin, H., 2002, 'Urban sprawl and public health', *Public Health Reports* May–June (117), 208.

Frumpkin, H., 2003, 'Healthy places: exploring the evidence', *American Journal of Public Health* 93, 1451–1456.

Hall, P., 1996, *Cities of Tomorrow*, Blackwell Publishers, Oxford.

Hall, P., 2009, 'Lecture at UC Berkeley College of Environmental Design', 25 September.

Hoekstra, A. Y., Chapagain, A. K., 2008, 'Globalization of water: sharing the planet's freshwater resources. The global component of freshwater demand and supply: an assessment of virtual water flows between nations as a result of trade in agricultural and industrial products', *Water International* 33 (1), 19–32.

Jackson, K., 1985, *Crabgrass Frontier: The Suburbanization of the United States*, Oxford Press, New York.

Lang, R., 2003, *Edgeless Cities*, Brookings Press, Washington, DC.

Lobell, D. B., Field, C. B., 2007, 'Global scale climate–crop yield relationships and the impacts of recent warming', *Environmental Research Letters* 2.

Mauer, C., 2005, 'Healthy nature healthy people: "contact with nature" as an upstream health promotion intervention for populations', *Health Promotion International* 21 (1), 48.

McKinsey Global Institute, 2009, *Preparing for China's Urban Billion*, Research Report (pp. 398–412) [available at www.mckinsey.com/mgi/publications/china_urban_summary_of_findings.asp].

Montgomery, D. R., 2007, *Dirt – The Erosion of Civilizations*, University of California Press, Berkeley, CA.

Morland, K., 2002, 'Neighborhood characteristics associated with the location of food stores and food service places', *American Journal of Preventive Medicine* 22 (1), 23–29.

Oatkiss, P., 2005, *The Validity of Food Miles as a Sustainability Indicator*, DEFRA Report, UK.

Ozer, E., 2007, 'The effects of school gardens on students andschools: conceptualization and considerations for maximizing healthy development', *Health Education Behavior* 34, 846–864.

Putnam, R., 2000, *Bowling Alone*, Touchstone, New York, 327–335.

Robinson, T., 1998, 'Does television cause childhood obesity', *Journal of the American Medical Association* 279, 959–960.

Sallis, J., Frank, L. D., Saelens, B. E., Kraft, M. K., 2004, 'Active transportation and physical activity: opportunities for collaboration on transportation and public health research', *Transportation Research A* 38, 249–268.

Sallis, J., Cervero, R. B., Ascher, W., Henderson, K. A., Kraft, M. K., Kerr, J., 2006, 'An ecological approach to creating active living communities', *Annual Review of Public Health* 27, 297–322.

Scott, M. M., Cohen, D. A., Evenson, K. R., Elder, J., Catellier, D., Ashwood, J. S., Overtona, A., 2007, 'Weekend schoolyard accessibility, physical activity, and obesity: the Trial of Activity in Adolescent Girls (TAAG) study', *Preventive Medicine* 44, 398–403.

Sokolov, A. P., Stone, P. H., Forest, C. E., Prinn, R., Sarofim, M. C., 2009, 'Probabilistic forecast for 21st century climate based on uncertainties in emissions (without policy) and climate parameters', *Journal of Climate* 169 [available at http://globalchange.mit.edu/files/document/MITJPSPGC_Rpt169.pdf].

Sturm, R., Cohen, D., 2004, 'Suburban sprawl and physical and mental health', *Public Health* 118, 488–496.

Takano, T., Nakamura, K., Watanabe, M., 2002, 'Urban residential environments and senior citizens' longevity in megacity areas: the importance of walkable green spaces', *Journal of Epidemiology and Community Health* 56, 913–918.

Thomas, J., Drukker, C., 2009, 'Returning to their roots', *Urban Land Institute Supplement* Spring, 43–47.

Ulrich, R., 1984, 'View through a window may influence recovery from surgery', *Science* 224, 420–421.

Ulrich, R. S., Simons, R. F., Losito, B. D., Fiorito, E., Miles, M.A., Zelson, M., 1991, 'Stress recovery during exposure to natural and urban environments', *Journal of Environmental Psychology* 11, 201–230.

United Nations, 2009, *Millennium Development Goals Report* (p. 47). United Nations, New York.

Viljoen, A., (ed.) 2005, *Continuous Productive Urban Landscapes. Designing Urban Agriculture for Sustainable Cities*, Architectural Press, London.

Williams, D., Collins, C., 2001, 'Racial residential segregation: a fundamental cause of racial disparities in health', *Public Health Reports* 116, 404–416.

Williams, R., 1975, *The Country and the City*, Oxford University Press, Oxford.

The Editors would like to thank the following people for their help in reviewing the papers for this special issue of the *International Journal of Agricultural Sustainability*:

Jill Auburn
CSREES, USDA
USA

Michael Bell
University of Wisconsin-Madison
USA

Nikita Eriksen-Hamel
Canadian International Development Agency
Canada

Cornelia Butler Flora
Iowa State University
USA

Rachel Hine
University of Essex
UK

Margaret Pasquini
Centro Interdisciplinario de Estudios sobre Desarrollo, Universidad de los Andes
Colombia

Mark Redwood
International Development Research Centre (IDRC)
Canada

Printed and bound by CPI Group (UK) Ltd, Croydon, CR0 4YY

22/10/2024

01777611-0015